VOLUME ONE HUNDRED AND SIXTY TWO

ADVANCES IN
CANCER RESEARCH
Redox Signaling

VOLUME ONE HUNDRED AND SIXTY TWO

ADVANCES IN
CANCER RESEARCH
Redox Signaling

Edited by

DANYELLE TOWNSEND
*Pharmaceutical and Biomedical Sciences,
Medical University of South Carolina, United States*

ED SCHMIDT
*Department of Microbiology and Cell Biology,
Montana State University, Bozeman, MT, United States;
Laboratory of Redox Biology, University of Veterinary Medicine
Budapest, Hungary*

ACADEMIC PRESS
An imprint of Elsevier

Academic Press is an imprint of Elsevier
50 Hampshire Street, 5th Floor, Cambridge, MA 02139, United States
525 B Street, Suite 1650, San Diego, CA 92101, United States
125 London Wall, London, EC2Y 5AS, United Kingdom

First edition 2024

Copyright © 2024 Elsevier Inc. All rights are reserved, including those for text and data mining, AI training, and similar technologies.

Publisher's note: Elsevier takes a neutral position with respect to territorial disputes or jurisdictional claims in its published content, including in maps and institutional affiliations.

No part of this publication may be reproduced or transmitted in any form or by any means, electronic or mechanical, including photocopying, recording, or any information storage and retrieval system, without permission in writing from the publisher. Details on how to seek permission, further information about the Publisher's permissions policies and our arrangements with organizations such as the Copyright Clearance Center and the Copyright Licensing Agency, can be found at our website: www.elsevier.com/permissions.

This book and the individual contributions contained in it are protected under copyright by the Publisher (other than as may be noted herein).

Notices
Knowledge and best practice in this field are constantly changing. As new research and experience broaden our understanding, changes in research methods, professional practices, or medical treatment may become necessary.

Practitioners and researchers must always rely on their own experience and knowledge in evaluating and using any information, methods, compounds, or experiments described herein. In using such information or methods they should be mindful of their own safety and the safety of others, including parties for whom they have a professional responsibility.

To the fullest extent of the law, neither the Publisher nor the authors, contributors, or editors, assume any liability for any injury and/or damage to persons or property as a matter of products liability, negligence or otherwise, or from any use or operation of any methods, products, instructions, or ideas contained in the material herein.

ISBN: 978-0-443-29444-0
ISSN: 0065-230X

For information on all Academic Press publications
visit our website at https://www.elsevier.com/books-and-journals

Publisher: Zoe Kruze
Editorial Project Manager: Naiza Ermin Mendoza
Production Project Manager: James Selvam
Cover Designer: Gopalakrishnan Venkatraman
Typeset by MPS Limited, India

Contents

Contributors   xi

1. Unresolved questions regarding cellular cysteine sources and their possible relationships to ferroptosis   1

Elias S.J. Arnér and Edward E. Schmidt

1. Roles and critical requirement of cellular cysteine	2
1.1 Sources of intracellular cysteine and modes of its entry into cells	5
1.2 GSH in cells and in circulation	11
1.3 GSSG in cells and in circulation	13
1.4 Other possible sources of cytosolic cys	14
1.5 Roles of NADPH, the NADPH-dependent disulfide reductase systems, and redoxins in CSSC reduction	14
2. Ferroptosis	16
2.1 Erastin	17
2.2 Protection/prevention against ferroptosis by iron-chelators	20
2.3 Association with lipid peroxidation	20
3. Role of GPX4 in ferroptosis	21
3.1 Kinetics and substrate specificities of GPX4	21
3.2 Results that question the exact mechanisms of GPX4 targeting in ferroptosis	22
4. FSP1 and other enzymes affecting ferroptosis in parallel with GPX4	26
4.1 FSP1	26
4.2 DHODH	27
4.3 MBOAT1 and MBOAT2	27
4.4 Proteins affecting iron status	27
5. Genetic and pharmacologic models that question the exact roles of cystine, Cys, xCT, GSH, or GSH synthesis in ferroptosis	28
5.1 CSSC and xCT	29
5.2 GSSG/GSH and GGT	30
5.3 GCL disruptions	30
5.4 GSR KO	31
6. Potential roles of the thioredoxin system in ferroptosis	31
7. Concluding remarks	36
Acknowledgments	37
References	37

2. Protein Tyrosine Phosphatase regulation by Reactive Oxygen Species 45

Colin L. Welsh and Lalima K. Madan

1. Introduction	46
2. The Protein Tyrosine Phosphatase (PTP) catalytic domain	47
3. A Cys-based catalytic mechanism and sensitivity to oxidation	50
4. Sources of Reactive Oxygen Species (ROS) in cellular signaling	53
5. Oxidation of PTPs I: Cellular mechanisms	56
6. Oxidation of PTPs II: Structural aspects	59
7. Conclusions	66
Acknowledgments	67
Disclosure of potential conflicts of interest	67
References	67

3. Mitochondrial metallopeptidase OMA1 in cancer 75

Gunjan Purohit, Polash Ghosh, and Oleh Khalimonchuk

1. Introduction	75
2. OMA1 and its role in mitochondrial and cellular physiology	78
3. Redox regulation of OMA1	82
4. OMA1 and its regulation in cancers	83
5. Clinical implications	89
6. Challenges and future directions	90
Acknowledgments	91
Conflict of interest	91
References	91

4. Role of antioxidants in modulating anti-tumor T cell immune resposne 99

Nathaniel Oberholtzer, Stephanie Mills, Shubham Mehta, Paramita Chakraborty, and Shikhar Mehrotra

1. Introduction	100
1.1 ROS production by T cells	100
1.2 Role of ROS in T cell signaling	101
1.3 Negative impact of ROS on T cell function	102
1.4 Antioxidant defense mechanisms in T cells	105
1.5 Sources of ROS in the TME	109
1.6 Antioxidant molecules and metabolites targeted in anti-tumor immunity	115

Contents   vii

     1.7 Clinical trials targeting immune cell antioxidant metabolism   115
     1.8 Conclusions and future directions   115
   Acknowledgment   116
   References   116

5. Redox pathways in melanoma   125

**Jie Zhang, Zhi-wei Ye, Danyelle M. Townsend, and
Kenneth D. Tew**

   1. Introduction   126
   2. Melanin biosynthesis   127
   3. MGST1 and melanoma   131
   4. MGST1 in melanoma metastasis and treatment response   132
   5. Additional redox targets in melanoma   135
   Acknowledgments   140
   References   140

6. Melanoma redox biology and the emergence of drug resistance   145

**Therese Featherston, Martina Paumann-Page, and
Mark B. Hampton**

   1. Biology of melanoma   147
   2. Treatment of melanoma   148
   3. Drug resistance in melanoma   149
   4. Oxidative stress in melanoma   152
   5. Antioxidants in melanoma   155
   6. Oxidative stress and BRAF inhibitor resistance   159
   Acknowledgments   163
   References   163

Contributors

Elias S.J. Arnér
Division of Biochemistry, Department of Medical Biochemistry and Biophysics, Karolinska Institutet, Stockholm, Sweden; Department of Selenoprotein Research and the National Tumor Biology Laboratory, National Institutes of Oncology, Budapest, Hungary

Paramita Chakraborty
Department of Surgery, Hollings Cancer Center, Medical University of South Carolina, Charleston, SC, United States

Therese Featherston
Mātai Hāora—Centre for Redox Biology and Medicine, Department of Pathology and Biomedical Science, University of Otago, Christchurch, New Zealand

Polash Ghosh
Department of Biochemistry, University of Nebraska-Lincoln, Lincoln, NE, United States

Mark B. Hampton
Mātai Hāora—Centre for Redox Biology and Medicine, Department of Pathology and Biomedical Science, University of Otago, Christchurch, New Zealand

Oleh Khalimonchuk
Department of Biochemistry, University of Nebraska-Lincoln; Nebraska Redox Biology Center, Lincoln; Fred & Pamela Buffett Cancer Center, Omaha, NE, United States

Lalima K. Madan
Department of Cell and Molecular Pharmacology & Experimental Therapeutics, College of Medicine; Hollings Cancer Center, Medical University of South Carolina, Charleston, SC, United States

Shikhar Mehrotra
Department of Surgery, Hollings Cancer Center, Medical University of South Carolina, Charleston, SC, United States

Shubham Mehta
Department of Surgery, Hollings Cancer Center, Medical University of South Carolina, Charleston, SC, United States

Stephanie Mills
Department of Surgery, Hollings Cancer Center, Medical University of South Carolina, Charleston, SC, United States

Nathaniel Oberholtzer
Department of Surgery, Hollings Cancer Center, Medical University of South Carolina, Charleston, SC, United States

Martina Paumann-Page
Mātai Hāora—Centre for Redox Biology and Medicine, Department of Pathology and Biomedical Science, University of Otago, Christchurch, New Zealand

Gunjan Purohit
Department of Biochemistry, University of Nebraska-Lincoln, Lincoln, NE, United States

Edward E. Schmidt
Department of Microbiology and Cell Biology, Montana State University, Bozeman, MT, United States; Laboratory of Redox Biology, University of Veterinary Medicine Budapest, Hungary

Kenneth D. Tew
Department of Cell and Molecular Pharmacology and Experimental Therapeutics, Medical University of South Carolina, Charleston, SC, United States

Danyelle M. Townsend
Department of Drug Discovery and Biomedical Sciences, Medical University of South Carolina, Charleston, SC, United States

Colin L. Welsh
Department of Cell and Molecular Pharmacology & Experimental Therapeutics, College of Medicine, Medical University of South Carolina, Charleston, SC, United States

Zhi-wei Ye
Department of Cell and Molecular Pharmacology and Experimental Therapeutics, Medical University of South Carolina, Charleston, SC, United States

Jie Zhang
Department of Cell and Molecular Pharmacology and Experimental Therapeutics, Medical University of South Carolina, Charleston, SC, United States

CHAPTER ONE

Unresolved questions regarding cellular cysteine sources and their possible relationships to ferroptosis

Elias S.J. Arnér[a,b] and Edward E. Schmidt[c,d,*]

[a]Division of Biochemistry, Department of Medical Biochemistry and Biophysics, Karolinska Institutet, Stockholm, Sweden
[b]Department of Selenoprotein Research and the National Tumor Biology Laboratory, National Institutes of Oncology, Budapest, Hungary
[c]Laboratory of Redox Biology, University of Veterinary Medicine, Budapest, Hungary
[d]Department of Microbiology and Cell Biology, Montana State University, Bozeman, MT, United States
*Corresponding author e-mail address: eschmidt@montana.edu

Contents

1. Roles and critical requirement of cellular cysteine	2
1.1 Sources of intracellular cysteine and modes of its entry into cells	5
1.2 GSH in cells and in circulation	11
1.3 GSSG in cells and in circulation	13
1.4 Other possible sources of cytosolic cys	14
1.5 Roles of NADPH, the NADPH-dependent disulfide reductase systems, and redoxins in CSSC reduction	14
2. Ferroptosis	16
2.1 Erastin	17
2.2 Protection/prevention against ferroptosis by iron-chelators	20
2.3 Association with lipid peroxidation	20
3. Role of GPX4 in ferroptosis	21
3.1 Kinetics and substrate specificities of GPX4	21
3.2 Results that question the exact mechanisms of GPX4 targeting in ferroptosis	22
4. FSP1 and other enzymes affecting ferroptosis in parallel with GPX4	26
4.1 FSP1	26
4.2 DHODH	27
4.3 MBOAT1 and MBOAT2	27
4.4 Proteins affecting iron status	27
5. Genetic and pharmacologic models that question the exact roles of cystine, Cys, xCT, GSH, or GSH synthesis in ferroptosis	28
5.1 CSSC and xCT	29
5.2 GSSG/GSH and GGT	30
5.3 GCL disruptions	30
5.4 GSR KO	31

Advances in Cancer Research, Volume 162
ISSN 0065-230X, https://doi.org/10.1016/bs.acr.2024.04.001
Copyright © 2024 Elsevier Inc. All rights are reserved, including those for text and data mining, AI training, and similar technologies.

6. Potential roles of the thioredoxin system in ferroptosis	31
7. Concluding remarks	36
Acknowledgments	37
References	37

Abstract

Cysteine is required for synthesis of glutathione (GSH), coenzyme A, other sulfur-containing metabolites, and most proteins. In most cells, cysteine comes from extracellular disulfide sources including cystine, glutathione-disulfide, and peptides. The thioredoxin reductase-1 (TrxR1)- or glutathione-disulfide reductase (GSR)-driven enzymatic systems can fuel cystine reduction via thioredoxins, glutaredoxins, or other thioredoxin-fold proteins. Free cystine enters cells thorough the cystine-glutamate antiporter, xCT, but systemically, plasma glutathione-disulfide might predominate as a cystine source. Erastin, inhibiting both xCT and voltage-dependent anion channels, induces ferroptotic cell death, so named because this type of cell death is antagonized by iron-chelators. Many cancer cells seem to be predisposed to ferroptosis, which has been proposed as a targetable cancer liability. Ferroptosis is associated with lipid peroxidation and loss of either glutathione peroxidase-4 (GPX4) or ferroptosis suppressor protein-1 (FSP1), which each prevent accumulation of lipid peroxides. It has been suggested that an xCT inhibition-induced cellular cysteine-deficiency lowers GSH levels, starving GPX4 for reducing power and allowing membrane lipid peroxides to accumulate, thereby causing ferroptosis. Aspects of ferroptosis are however not fully understood and need to be further scrutinized, for example that neither disruption of GSH synthesis, loss of GSH, nor disruption of glutathione disulfide reductase (GSR), triggers ferroptosis in animal models. Here we reevaluate the relationships between Erastin, xCT, GPX4, cellular cysteine and GSH, RSL3 or ML162, and ferroptosis. We conclude that, whereas both Cys and ferroptosis are potential liabilities in cancer, their relationship to each other remains insufficiently understood.

1. Roles and critical requirement of cellular cysteine

Although the abundance of cysteine (Cys) is 20th out of the 21 translated amino acids in the human proteome, with only the hyperreactive selenium (Se)-containing Cys-analog selenocysteine (Sec) having lower representation (Sec occurs in 25 human proteins, <0.1% of the proteome); 92% of human proteins contain at least one Cys and it represents 2.26% of the amino acids in the human proteome (Miseta & Csutora, 2000). Cys is also used for biosynthesis of glutathione (γ-L-glutamyl-L-cysteinyl glycine; GSH), coenzyme A (CoA), H_2S, taurine, iron–sulfur clusters, and other sulfur (S)-containing metabolites. GSH, the most abundant intracellular low molecular weight thiol (3-10 mM) (Hansen, Roth, & Winther, 2009), is noteworthy as its thiol-S comes exclusively from Cys (Stipanuk, Dominy,

Lee, & Coloso, 2006). No cell could likely survive without a reliable source of Cys. Cys itself, however, is unstable in the extracellular environment, wherein it effectively oxidizes into the disulfide-linked dimeric form of Cys, cystine (hereafter abbreviated CSSC; see below). Thus, cells typically acquire Cys by assimilating CSSC from extracellular sources and reducing this to 2 Cys in the cytosol, although cancer cells can often upregulate alternative sources of Cys, such as more *de novo* synthesis through transsulfuration due to activation of transcription factor "nuclear factor, erythroid-derived-2, like-2" (NRF2) in cancer cells (Bonifacio, Pereira, Serpa, & Vicente, 2021); the transsulfuration pathways is discussed at further detail below.

Cellular dependence upon Cys is considered a potentially targetable liability in cancer (Combs & DeNicola, 2019; Daher, Vucetic, & Pouyssegur, 2020; Jyotsana, Ta, & DelGiorno, 2022). As with any therapy, to be useful, cancer cells need to be substantially more sensitive to Cys-targeting therapies than other cells in the body. Moreover, it is important to understand mechanistically how Cys-targeting impacts cell physiology systemically, such that its targeting for cancer therapy can be effectively directed toward having more potent impact on the cancer cells than on other cells in the patient.

Metabolic properties of many cancer cells result in their having chronically elevated levels of reactive oxygen species (ROS). It is logical that this is a liability that might be targetable for therapy (Trachootham, Alexandre, & Huang, 2009). Cellular ROS in the form of peroxides (ROOH) are detoxified in mammalian cells predominantly by the two families of abundant peroxidases, the GSH peroxidases (GPXs) and the peroxiredoxins (PRXs) (Winterbourn, 2013). These, in turn, obtain electrons to drive their reduction reactions from reduced nicotinamide adenine dinucleotide phosphate (NADPH), either via GSH and glutathione disulfide reductase (GSR; NADPH→GSR→GSH→GPX→ROOH) or via thioredoxin-1 (Trx1) and thioredoxin disulfide reductase (TrxR1; NADPH→TrxR1→TRX1→PRX→ROOH) (Winterbourn, 2013). Because GSH links NADPH with GPXs, and Cys is required for GSH synthesis, it has been proposed that limiting Cys will lower cellular GSH levels, thereby indirectly impeding GPX activity and increasing ROS levels to yield oxidative stress and cell death (Jyotsana et al., 2022; Rosell et al., 2023; Stockwell & Jiang, 2020; Xu et al., 2019). Cancer cells generating increased levels of ROS, therefore, should be particularly susceptible to diminished peroxidase activity. Because ferroptosis is associated with accumulation of phopholipid hydroperoxides (discussed further in Section 1.4) and, of all the peroxidases in either family, only GPX4 shows substantial reduction of phospholipid hydroperoxides (Brigelius-Flohe, 2006; Flohe, Toppo, Cozza, & Ursini, 2011; Schwarz et al., 2023), it seems reasonable that lowering the GSH levels might be particularly effective at promoting ferroptosis even if the NADPH→TrxR1→Trx1→PRX system remains intact. However, several observations suggest that this notion may be too simplified, as we shall further discuss herein.

The unique attribute of Cys in the proteome is the free thiol-S (-SH) in its side chain. Methionine (Met), the other proteinogenic S-amino acid, has a thioether-S which, although being at the same redox state as a thiol, is much less reactive with oxidants compared with Cys (Kaya, Lee, & Gladyshev, 2015). This allows Met to remain reduced extracellularly also when ingested as a nutrient in food, or when present in proteins or blood plasma. However, Met may nonetheless become oxidized into its sulfoxide derivatives, which in turn can be repaired by TrxR1/Trx1-dependent methionine sulfoxide reductases (Kim & Gladyshev, 2007; Le et al., 2008). Cys being more prone to oxidation than Met, is less utilized in proteins (Miseta & Csutora, 2000) and is typically found only in positions that require the unique chemistry of the thiol: within *i*) active sites, *ii*) regulatory sites that are influenced by derivatization of the thiol, or *iii*) in structural positions where they form intra- or inter-molecular disulfide bonds (Fomenko, Marino, & Gladyshev, 2008).

In the reducing environment of the cytosol, nearly all Cys residues will be in the reduced thiol state. Whereas the Cys-thiol is thermodynamically susceptible to oxidation by biological oxidants like superoxide radical (O_2^-) or H_2O_2, the reaction rates for these oxidants with "typical cellular Cys-thiols" is between 10^0 and $10^1 \cdot mol^{-1} \cdot sec^{-1}$ (Winterbourn & Hampton, 2008; Winterbourn & Metodiewa, 1999; Winterbourn, 2013). Whereas this rate is sufficient to effectively oxidize most thiols to disulfides in the oxidizing environment of blood plasma, these rates are too slow for contributing to intracellular processes (Nagy et al., 2011; Winterbourn & Hampton, 2008; Winterbourn, 2013). "Typical" cellular Cys-thiols include free Cys, GSH, and nearly all protein-Cys-thiols; indeed the total intracellular protein-derived free thiol concentrations has been estimated to be significantly higher than that derived from GSH (Requejo, Hurd, Costa, & Murphy, 2010). Such thiol groups on proteins can clearly participate in redox reactions and thus contribute to the total antioxidant defense capacity of the cells (Requejo et al., 2010). In addition, the PRX and GPX families of "professional peroxidase" proteins exhibit active-site Cys (PRXs) or Sec (most GPXs) residues with $\sim 10^7$-fold more reactivity towards their peroxide substrates than that seen with typical non-catalytic thiols (Nagy et al., 2011; Winterbourn, 2013). In the case of PRX, the H_2O_2-reactive Cys, termed the "peroxidatic Cys", achieves its high rate through a specialized microenvironment in the active site that both activates the Cys by lowering its pKa, resulting in a more reactive nucleophilc "thiolate anion", and activates the enzyme-bound H_2O_2 molecule for facilitated catalysis (Nagy et al., 2011). GPXs likely use similar active site specialization, along with the higher intrinsic reactivity of the Se of Sec versus the S atom of Cys, combined with additional enzyme features to achieve very high rates of reactivity with their peroxide substrates (Cozza et al., 2017; Toppo, Flohe, Ursini, Vanin, & Maiorino, 2009). Finally, a small number of other proteins, including glyceraldehyde-3 phosphate dehydrogenase (GAPDH) and protein-tyrosine phosphatase 1B (PTP1B), have specialized microenvironment contexts that similarly increase their reactivity with H_2O_2 by several orders of magnitude, making these also directly reactive with H_2O_2 at physiologically significant rates and concentrations (Dagnell et al., 2017; Doka et al., 2020; Netto & Machado, 2022; Talwar et al., 2023).

1.1 Sources of intracellular cysteine and modes of its entry into cells

1.1.1 De novo Cys synthesis

Many plants, fungi, eubacteria, and archaea reduce inorganic sulfur species into sulfide, which is then used for primary synthesis of Cys-thiol from serine (Ser) (Liu et al., 2021b; Russel, Model, & Holmgren, 1990; Wray, Campbell, Roberts, & Gutierrez-Marcos, 1998). Most if not all plants, fungi, eubacteria, and archaea are also able to use the thiol-S from Cys for de novo synthesis of Met (Miller & Schmidt, 2019). Perhaps due to the strictly heterotrophic life-history of metazoans providing an adequate source of protein-Cys, metazoans lack the ability to reduce inorganic sulfur, or to convert Cys to Met. Instead, the Cys-to-Met pathway was effectively reversed at the metazoan transition, exchanging the ability to synthesize Met from Cys for the ability to synthesize Cys de novo using Met as the S-source (Miller & Schmidt, 2019; Miller & Schmidt, 2020). As such, Met is an essential amino acid in metazoans.

De novo synthesis of Cys from Ser + Met in metazoans is a 5-step process in which Met-derived homocysteine (Hcy) is shunted-off of the Met cycle into the transsulfuration pathway (Fig. 1A). The enzyme cystathionine-β-synthase (CBS) covalently ligates Hcy to Ser by elimination of the Ser sidechain-hydroxyl and formation of a C-S bond with the Hcy-thiol, resulting in the thioether

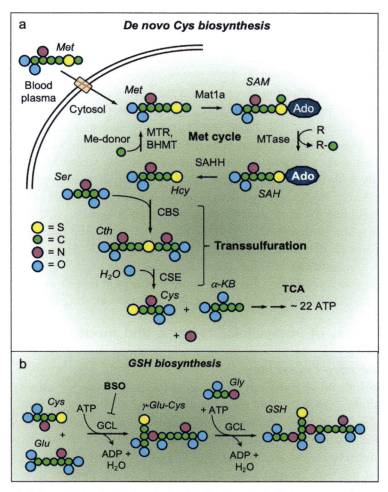

Fig. 1 **Cellular synthesis of Cys and GSH.** (A) Atomic ball diagram of de novo Cys synthesis. Splanchnic organs, in particular liver, synthesize Cys de novo using the S from Met, an essential amino acid, and the amino acid backbone from Ser. In this 5-step pathway Met transits the Met-cycle to Hcy, whereat rather than being re-methylated to Met, it is shunted to transsulfuration. Only the S from Met appears in Cys; most of the Met amino acid backbone eventually enters the TCA cycle where its oxidation generates an estimated 22 ATP phosphodiester bonds. Abbreviations not in text: Ado, adenosine; α-KB, α-ketobutyrate; BHMT, betaine hydroxymethyltransferase; MAT1a, Met-adenosyltransferase-1a; Me, methyl; MTase, SAM-dependent methyl-transferases; MTR, 5-methyltetrahydrofolate-homocysteine methyltransferase; SAH, S-adenosylhomocysteine; SAHH, SAH-hydrolase; SAM, S-adenosylmethionine; (B) Atomic ball diagram of de novo GSH synthesis. Both Cys and GSH biosynthesis are strictly cytosolic.

cystathionine (Cth). In a second step, Cth is cleaved at the other C-S bond by cystathionine-γ-lyase (CSE, cystathionase), resulting in Cys, α-ketobutyrate, and NH$_3$ (Miller, Holmgren, Arner, & Schmidt, 2018). Transsulfuration consumes no NADPH and requires no disulfide reduction step (Miller et al., 2018). Only the S atom of Met is passed on to the newly synthesized Cys. Whereas the pathway consumes Met, Ser, and 3 phosphodiester bonds from ATP, the Met-derived α-KB is metabolized into succinyl-CoA, whose subsequent oxidation in the tricarboxylic acid (TCA) cycle is predicted to generate 22 ATP phospho-diester bonds (Eriksson, Prigge, Talago, Arner, & Schmidt, 2015), suggesting that if Met and Ser are not limiting, this pathway is likely to be energetically favorable (Fig. 1A). Moreover, this uniquely metazoan pathway provides an NADPH-independent source of Cys, which contributes to long-term survival in mouse livers having genetic disruptions in the disulfide reducing systems (Miller et al., 2018); shown to contribute to the sustained long-term homeostasis in mouse livers lacking both TrxR1 and GSR (Eriksson et al., 2015). It is also thought to be important in supporting hepatocyte survival under severe oxi-dative stress or upon toxic insults that either deplete NADPH or compromise TrxR1 and GSR in the liver (Miller & Schmidt, 2020). However, whereas transsulfuation in hepatocytes is supported by an active Met-cycle including the specialized liver-specific Met-adenosyltransferase-1a enzyme (MAT1a), and provides an estimated 30–50% of the Cys in liver, few other cell types, unless stressed, express substantial levels of both CBS and CSE (Finkelstein, 1998; Kabil, Vitvitsky, Xie, & Banerjee, 2011). A predicted rational for transsulfuration in the liver and, to a lesser extent, in kidney, pancreas, and small intestine, is that these splanchnic organs support intermediary metabolism of Cys (Finkelstein, 1998). The ability to use either dietary disulfides or Met + Ser in order to generate Cys might help ensure reliable systemic Cys availability. Interestingly, it has been reported that cancers arising from normally transsulfuration-proficient cells (e.g., liver, pancreas, kidney) lose the expression of one or both transsul-furation enzymes during transformation, making reliance on CSSC reduction a potential acquired liability (Combs & DeNicola, 2019). Irrespective of the source of Cys in the cytosol, it will be readily available for GSH synthesis in a two-step pathway requiring glutamate, glycine and ATP (Fig. 1B).

1.1.2 Cys in cells and in circulation

Food proteins provide the dietary source of sulfur amino acids. Met is essential (must be provided by diet); Cys must either be provided from the diet or synthesized *de novo* using S derived from dietary Met (see above). Following digestion to free amino acids or (sub-antigenic) di- and oligopeptides in the gut lumen, with subsequent absorption by the small intestine, these nutrients enter the enterohepatic circulation (Trommelen, Tomé, & van Loon, 2021). First-pass absorption goes from the enterohepatic circulation to splanchnic tissues (small intestine, liver, and pancreas); systemic circulating amino acids and peptides have either escaped first-pass absorption, or were re-introduced to the circulation by the splanchnic tissues, in particular liver, for use in intermediary metabolism (Trommelen et al., 2021).

For Cys to be useful in intermediary metabolism, several criteria must be met: (*i*) The form of Cys present in circulation needs to be in a thiol- or disulfide-state; higher oxidation states cannot be used as amino acid sources in mammals (Miller & Schmidt, 2020); (*ii*) cells exporting Cys to circulation need to have a mechanism of producing and exporting it in that form (thiol- or disulfide-state); and (*iii*) A recipient cell requiring Cys needs to have a mechanism of assimilating Cys from circulation as well as recovering

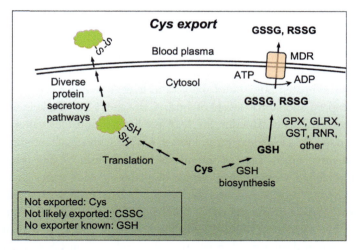

Fig. 2 Cellular Cys export routes. Diagram shows established routes of Cys export from cells. Although CSSC can, in principal, be exported by xCT in exchange for extracellular Glu, there is no known mechanism for producing CSSC in the cytosol and the exceptionally low CSSC concentration in cytosol favors CSSC import by xCT. Also, the cytotoxicity associated with disrupted Cys-catabolism indicates cells cannot likely excrete Cys nor CSSC. GSSG is exported by MDRs in ATP-dependent reactions. Although GSSG and GSSR are very low abundance in cytosol, they are catalytically generated by GPX, GLRX, GST, RNR, and other enzymes, which might drive export. GSH exists extracellularly but this could be derived from thiol-disulfide exchange between GSSG and plasma protein thiols; to date a GSH exporter has not been validated.

the free Cys in its reduced thiol form, in order to make use of it. To date, only few routes are known to meet these criteria for Cys export (Fig. 2).

The most well-established mechanisms of exporting Cys from cells are via secretion of Cys-containing proteins through secretory pathways, which can then be assimilated by pinocytosis, and via secretion of GSSG or other GSH-conjugates by ATP-binding cassette (ABC) exporters, in particular the multidrug resistance protein-1 (MDR1) (Oestreicher & Morgan, 2019).

CSSC is also found in blood plasma but its sources remain unclear. In principle, CSSC could be secreted by hepatocytes via xCT in exchange for extracellular Glu, especially if extracellular Glu and intracellular CSSC are both relatively high. However, this seems unlikely because cytosolic CSSC levels are exceptionally low in normal cells (Kabil et al., 2011), thereby likely favoring xCT to import CSSC and export Glu (Fig. 3A). Recently, a dedicated CSSC exporter, cystinosin, has been reported allowing secretion of CSSC from the lumen of intracellular vesicles into the cytosol, e.g. following pinocytosis or autophagy (Fig. 3B); however cystinosin is not reported to support extracellular excretion (Guo et al., 2022). Moreover, there is no known catalytic mechanism to generate CSSC from cytosolic Cys to allow for its excretion. Theoretically, the more oxidizing environment in the lumen of the endoplasmic reticulum (ER) might support an ER-mediated route of CSSC excretion (Reznik & Fass, 2022); however, we remain unaware of evidence supporting this route, as yet. Based on current knowledge, plasma CSSC appears to predominantly arise directly from intestinal protein digestion.

It has been proposed that free Cys in plasma can be assimilated into cells via neutral-amino acid transporters and perhaps SLC1A1 (Combs & DeNicola, 2019), although evidence supporting either yet remains unclear (Fig. 3A). Regardless, because Cys is neither stable nor abundant in plasma, such direct uptake likely contributes insignificantly to cellular Cys pools. Of interest, the transporter SLC1A5, originally named "Ala, Ser, and Cys Transporter-2" (ASCT2) is now known to be a Na^+-dependent antiporter that imports Gln in exchange for Ser, Asn, or Thr; Cys is not a substrate but increased intracellular Cys modulates ASCT2, causing it to, instead, export Gln (Scalise, Pochini, Console, Losso, & Indiveri, 2018). The instability and low concentration of Cys in extracellular fluids is due to the oxidizing plasma environment, which favors oxidation of Cys into either CSSC or mixed disulfides; the flavoenzyme quiescin-sulfhydryl oxidase 1 (QSOX1, ~25 nM in human serum) also efficiently catalyzes oxidation of plasma thiols to disulfides with the coincident reduction of molecular oxygen to H_2O_2

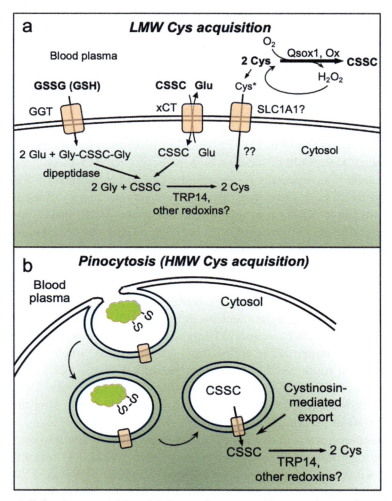

Fig. 3 Cellular Cys acquisition. (A) Cys can be acquired by assimilation of circulating GSSG or CSSC via GGT or xCT, respectively. Extracellular GSH can also be assimilated by GGT and extracellular Cys might be assimilated by neutral amino acid transporters or SLC1A1, when present. However Cys and other thiols are low-abundance in the plasma due to chemical oxidation and enzymatic oxidation via Qsox1 which, in the case of Cys, generate CSSC. Asterisk, Cys oxidation in the extracellular environment casts doubt on the physiological availability of extracellular Cys; question marks, uncertain activities. (B) Plasma proteins can be assimilated by pinocytosis. Following proteolysis to amino acids in the lysosome, the CSSC is exported into the cytosol by cystinosin. Regardless the route of entry, CSSC needs to be converted to Cys intracellularly. TRP14, fueled by NADPH-TrxR1, appears to be the predominant CSSC reductase. (A) Assimilation of LMW sources. (B) Assimilation of HMW sources.

(Fig. 3A) (Israel, Jiang, Gannon, & Thorpe, 2014). Moreover, although dietary sources might release some reduced Cys into the enterohepatic circulation, there is no means for cells to excrete reduced Cys into plasma or interstitial fluids to support its use in intermediary metabolism. No Cys exporter is known, and the cytotoxic phenotype associated with livers or cells having Cys catabolism deficiencies indicate that Cys, itself, cannot be exported from cells. Rather, elimination of excess intracellular Cys proceeds by Cys-dioxygenase (CDO)-catalyzed oxidation to Cys-sulfinate, a non-reversible oxidation state, followed by metabolism to hypotaurine (Htau) and then taurine (Tau) prior to final excretion (Stipanuk & Ueki, 2011; Stipanuk et al., 2006). Thus, it evident that tissues contributing to Cys intermediary metabolism excrete forms other than free Cys (for more discussions on this topic, see below).

1.2 GSH in cells and in circulation

GSH is found across all phyla; in most cells it is the most abundant low molecular weight (LMW) thiol, present in the cytosol at a steady state concentration of 3–10 mM (Oestreicher & Morgan, 2019; Stenersen, Kobro, Bjerke, & Arend, 1987; Thuillier et al., 2011). By contrast, Cys concentrations are only 20–100 μM (Stipanuk et al., 2006), i.e. ~0.2–3% that of GSH. Neither GSH nor Cys are particularly reactive with cellular oxidants in direct reactions (see above) (Winterbourn & Metodiewa, 1999); however, unlike free Cys, GSH can support GPXs to drive reduction of peroxides (Brigelius-Flohe & Maiorino, 2013; Toppo et al., 2009). Thus, whereas it is unlikely that Cys could replace GSH for protecting cells against oxidative stress, it is plausible that Cys restriction might lead to secondary GSH depletion. Consistent with this, it has been suggested that cytosolic Cys depletion, by leading to GSH depletion, hinders the flow of reducing power via GSH to GPXs (Chiang et al., 2022).

One of the most surprising findings about GSH came with the discovery nearly 50 years ago of the drug buthionine sulfoximine (BSO), which effectively inhibits Glu-Cys ligase (GCL), the first committed and rate-limiting step in GSH biosynthesis (Fig. 1B) (Griffith & Meister, 1979b; Lu, 2013). BSO treatment of cells results in a precipitous drop in GSH levels, often to a few percent of the normal level; importantly, usually without causing cell death (Griffith & Meister, 1979a). Indeed, long-term high–dose BSO treatment in mice results in severe systemic GSH depletion throughout most organs, yet is well tolerated (Griffith & Meister, 1979a). Also, BSO is approved for use in human patients, wherein it has been used over the past 40 years (Green et al., 1984). More recently, genetically engineered mouse models were developed with either attenuated or disrupted GCL activity in specific cell types. Consistent with the mild consequences of pharmacologic inhibition of GCL by BSO, however, reported phenotypes do not include catastrophic cell death (Botta et al., 2008; Franklin et al., 2009; Harris et al., 2015; Kurniawan et al., 2020; Nakamura et al., 2011). Much clearly remains to be understood about how diverse cells as well as complex multicellular organisms tolerate the loss of this normally abundant thiol—nearly universally, GSH depletion is benign. This should be considered when evaluating models wherein a modest disruption in synthesis rates or slightly lowered steady state levels of GSH might induce oxidative stress or cell death (Lee & Roh, 2022).

In the cytosol, the GSH:GSSG ratio is highly favored towards the reduced state and has been estimated to be between 100:1 and 1000:1 (Morgan et al., 2013), or even as much as up to 40,000:1 (Morgan, Sobotta, & Dick, 2011); by contrast, in blood plasma the ratio is ~20:1 (Jones et al., 2000). We are aware of no reports of blood plasma having measurable levels of NADPH, an NADPH-generating system, nor a primary NADPH-dependent disulfide reductase system. Thus, plasma thiols must arise either from cytosolic disulfide reducing power or from primary protein digestion in the gut (see above). In combination, this suggests that cells can excrete reduced GSH; however, a mechanism for this remains to be found (Fig. 2). Previous reports of organic-anion transporter proteins (OATP), MDR1, or other transporters excreting GSH, have more recently been experimentally unsupported or refuted (Mahagita, Grassl, Piyachaturawat, & Ballatori, 2007). Until a GSH exporter is validated, the only known sources of plasma thiols are thus either dying cells or other mechanisms bluntly releasing all or parts of their cytosolic contents, secreted proteins with reduced Cys, or from primary digestion of protein in the gut. However, we should not exclude that a dedicated GSH exporter might exist and be identified in the future. Indeed, although the machinery for GSH synthesis exists only in the cell cytosol, GSH is found also in organelles, including mitochondria. Only very recently has a putative transporter been identified for moving GSH into the mitochondria (Liu et al., 2023). Inside the mitochondria, mitochondrial GSR recycles GSSG within the mitochondrial matrix independent of the cytosolic GSH/GSSG pool (Kelner & Montoya, 2000).

Extracellular GSSG can undergo disulfide-thiol exchange with plasma thiols—in particular with protein thiols on secreted proteins—thereby liberating one reduced GSH moiety while also glutathionylating the protein thiol with which it may have reacted. As such, extracellular GSH might arise not from GSH excretion, but from GSSG excretion followed by extracellular disulfide-thiol exchange. By this mechanism, however, the cytosol of the cell exporting a protein with a reduced Cys must have supplied the reducing power for converting GSSG + protein-SH into protein-SSG + GSH. Thus, disulfide-thiol exchange reactions like this cannot generate a net increase in free thiols, but can move the reduced thiol group between cells and onto a different molecular species.

1.3 GSSG in cells and in circulation

GSSG is found in blood plasma and, importantly, there are known routes of export of GSSG from hepatocytes. Like CSSC, GSSG is not abundant in cytosol (at least 1:100–1:1000 ratio for GSH:GSSG, see above); however unlike for CSSC, hepatocytes have numerous efficient enzymes for oxidizing reduced GSH into either GSSG or different mixed disulfide glutathionylated molecular species (GSSR). These enzymes include glutaredoxins (GLRX), GPXs, glutathione-S-transferases (GSTs), ribonucleotide reductase (RNR), and others (Cassier-Chauvat, Marceau, Farci, Ouchane, & Chauvat, 2023). The production of disulfides formed upon the catalytic cycles of these enzymes will be counteracted by the disulfide reductase systems; yet, having catalytic machineries producing GSSG or GSSR will, in contrast to the case for CSSC, make it plausible that these products can also be excreted from cells as a result of their continuous production (Fig. 2). Furthermore, unlike for either GSH or CSSC, there are well-characterized effective exporters for moving either GSSG or GSSR to the blood plasma. Several Multi-Drug Resistance (MDR, also called ATP-Binding Cassette, ABC) exporters secrete glutathionylated substrates out of cells (Van Der Kolk, Vellenga, Muller, & de Vries, 1999). Whereas their substrate specificities are low, one key excreted conjugate is GSSG (Fig. 2) (Ballatori, Krance, Marchan, & Hammond, 2009; Cole & Deeley, 2006). Indeed, effective export of GSSG might contribute to the maintenance of a highly reduced steady-state ratio of GSH:GSSG in cells or in subcellular compartments also under conditions wherein disulfide reduction is compromised (Eriksson et al., 2015; Kojer et al., 2012; Miller & Schmidt, 2020; Morgan et al., 2013).

There is also a well characterized mechanism for cells to acquire Cys from circulating GSSG or GSH (Fig. 3A). The ubiquitous outer membrane protein γ-glutamyl-transpeptidase (GGT) takes GSSG or GSH from the blood plasma, cleaves the γ-glutamyl bond and thereby generates, in the case of GSSG, 2 Glu + diglycinylcystine (Gly-CSSC-Gly). The 2 Glu and the

Gly-CSSC-Gly are then imported into the cell, wherein the peptide bonds between Gly and CSSC are cleaved by cytosolic dipeptidases (Kobayashi, Ikeda, Shigeno, Konno, & Fujii, 2020; Tate & Meister, 1981). The net result is import of CSSC, 2 Gly and 2 Glu; the CSSC is then reduced by the cytosolic disulfide reductase systems to give 2 Cys. In the case of GSH, the products acquired by the cell are instead Glu, Gly, and Cys. Of note, unlike the situation for xCT, which must export Glu to obtain CSSC, GGT imports CSSC, 2 Glu and 2 Gly, thus providing all three amino acids that are needed for GSH biosynthesis in the cell (Anderson & Meister, 1980; Deneke & Fanburg, 1989; Griffith & Meister, 1979a; Griffith, Bridges, & Meister, 1981; Meister, Griffith, Novogrodsky, & Tate, 1979). Hence, GSSG appears to be a favorable molecule for mediating the intermediary metabolism of Cys during export, import, and utilization (Fig. 3A).

1.4 Other possible sources of cytosolic cys

A recent study on pancreatic ductal adenocarcinoma suggested that non-cancer cells in the tumor microenvironment (TME) might directly transfer Cys or CSSC to the cancer cell cytosol in a cell-to-cell transfer reaction (Meira et al., 2021). The mechanisms, cell-type compatibility, and proportional contributions of this system remain to be resolved, but it would be consistent with cytosolic exchange between cancer and non-cancer TME cells, e.g., via desmosomes or microvesicle-mediated intercellular transport.

1.5 Roles of NADPH, the NADPH-dependent disulfide reductase systems, and redoxins in CSSC reduction

NADPH is the "universal currency" for anabolic reduction reactions (Miller et al., 2018). For disulfide reduction, including the reduction of CSSC → 2 Cys, nearly all reducing power is transited to the process from NADPH by one of mammalian cells only two families of NADPH-dependent disulfide reductases—the GSR and TrxR families (Fig. 4) (Arner & Holmgren, 2000; Holmgren, 2000; Miller et al., 2018). Cells and mouse model systems genetically lacking either GSR or TrxR1 are viable and robust (Bondareva et al., 2007; Jakupoglu et al., 2005; Mandal et al., 2010; Rogers, Tamura, Rogers, Welty, & Hansen, 2004; Suvorova et al., 2009), and TrxR1 inhibitors, including auranofin (AFN) and the more specific TrxR1-inhibitor-1 (TRi-1) are well-tolerated by healthy cells, animals, and for AFN humans (Arnér, 2009b; Sabatier et al., 2021; Stafford et al., 2018), indicating that loss of either TrxR1 or GSR does not cause either a Cys-insufficiency nor catastrophic cell death in any of these situations.

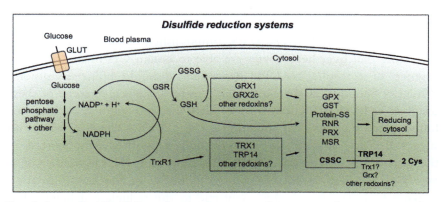

Fig. 4 Cytosolic disulfide reductase systems. Generation of cytosolic disulfide reducing power uses NADPH, generated largely from oxidation of glucose in the Pentose Phosphate Pathway. TrxR1 and GSR are the only two enzymes that can use NADPH to drive reduction of a disulfide bond. TrxR1 reduces Trx1-, TRP14- and other cytosolic TRX-disulfides; GSR reduces GSSG. GSH and reduced redoxins sustain the chemically reducing status of the cytosol and drive specific other reduction reactions. Redoxins reduce CSSC → 2 Cys, with TRP14 being the predominant CSSC reductase.

Neither GSR nor TrxR1 can directly reduce CSSC (Arner & Holmgren, 2000; Arner & Holmgren, 2006; Arnér, 2009b). Instead, reducing power must be transferred from these "primary reductases" via their downstream redoxin-family members of Trx-fold proteins (GLRXs, TRXs, or more distantly related family members) that finally would be able to reduce the CSSC disulfide bond. The details of how CSSC is reduced in cells remains unclear; however not all redoxins show equal reactivity with CSSC (Fig. 4). The TRX-related protein of 14 kDa (TRP14, encoded by *TXNDC17*), although unable to reduce disulfides in many protein-disulfides that are substrates for TRX1, shows at least a 5-fold greater reactivity with CSSC than does TRX1 (Eriksson et al., 2015; Pader et al., 2014). TRP14 is ubiquitously expressed and conserved across phyla, suggesting that TRP14 might be the predominant CSSC reductase in vivo, which we have recently confirmed (Marti-Andres, 2024). Notwithstanding, it is noteworthy that TRP14-null cells and mice are viable (Doka et al., 2016; Doka et al., 2020), showing that loss of TRP14 causes neither Cys-insufficiency nor catastrophic cell death. Clearly there must be other ubiquitous processes that cells can use to obtain sufficient Cys in the absence of TRP14. Even more dramatically, we have found that livers lacking both TrxR1 and GSR "TR/GR-null") (Eriksson et al., 2015; McLoughlin et al., 2019), TRX1/TR/GR-null livers (McLoughlin et al., 2019; Prigge et al., 2017), or even TRP14/TRX1/TR/GR-null livers (EES, unpublished data) remain viable, indicating that liver hepatocytes are able to obtain Cys in the complete absence of cytosolic NADPH-dependent disulfide reductase systems (Miller & Schmidt, 2019). Sustained hepatic Cys levels in some of these disulfide reductase-deficient liver models was also verified by metabolomic measurements (McLoughlin et al., 2019). These observations suggest that *de novo* synthesis of Cys can fully support liver homeostasis in the absence of CSSC reduction. Further supporting the importance of *de novo* Cys synthesis in these systems, inhibition of the transsulfuration enzyme CSE becomes highly cytotoxic to the reductase-deficient, but not to healthy wildtype livers (Eriksson et al., 2015). These diverse reductase-compromised mouse and cell models show that some, or most, normal and healthy cell types can thrive under conditions of limited or even fully inhibited CSSC reduction capacity (Eriksson et al., 2015; Miller & Schmidt, 2020; Pader et al., 2014).

2. Ferroptosis

In 2003, as a part of a large screen for small molecules that would kill cancer cells but not normal control cells, a novel compound named "Erastin" (for Eradicator of RAS and small-T antigen-expressing cells) was identified that induced a non-apoptotic type of cell death, particularly in cells that express mutant HRAS and small-T antigen (Dolma, Lessnick, Hahn, & Stockwell, 2003). In 2012, it was found that Erastin-induced cell death in HRAS mutant cancer cells could be antagonized by iron-chelators (Dixon et al., 2012). Based on the iron-association, this form of cell death, which was also found to be related to lipid peroxidation, was entitled "ferroptosis" (Dixon et al., 2012). Experimentally, ferroptosis can thus be defined as the type of cell death that is prevented by iron chelators (e.g., deferiprone, deferoxamine) or small lipophilic antioxidants (e.g., ferrostatin, liproxstatin) (Jiang, Stockwell, & Conrad, 2021; Yan et al., 2021).

In the following sections we will discuss the particular features and observations relating to experimental ferroptosis studies, relate these findings to the metabolism of Cys and GSH, also considering cytosolic reducing enzymes, and we will suggest where further scrutiny might be warranted due to a number of yet outstanding questions with regards to the mechanisms involved (Fig. 5).

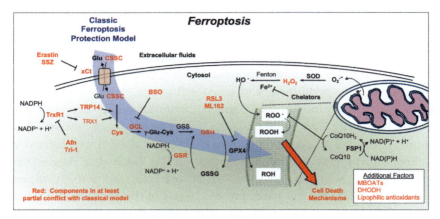

Fig. 5 Ferroptosis mechanisms and outstanding questions remaining to be answered. With ferroptosis defined as a cell death triggered by iron-dependent lipid peroxidation, the roles of Fe^{2+}, iron-chelators, $HO\cdot^-$, and lipid-radicals/lipid peroxidation are clear, as, very likely, is the role of GPX4, as a lipid-peroxidase (center of diagram) and FSP1 + CoQ10 shown to similarly repair lipid radical species. However, because ferroptosis was identified based on data from treating cultured cells with small molecules, including erastin or SSZ, and described as being a non-apoptotic cell death, the classical model of cellular protection against ferroptosis (blue arrow) has many components for which conflicting experimental findings are now known (components in red font, and likely others). **Table 1** outlines some of the outstanding questions related to these components.

2.1 Erastin

Erastin (PubChemID: 11214940) is a quinazoline compound that inhibits mitochondrial voltage gated anion channels (VDACs) and was subsequently proposed as an irreversible inhibitor of xCT. In the latter case, brief exposure of cells to erastin was sufficient to see a strong inhibitory effect on CSSC uptake, although it could not be demonstrated if or how erastin directly might have modified any component of xCT (Sato et al., 2018). Interestingly, a more recent study suggested that erastin's inhibition of VDACs directly leads to oxidative stress and cell death, without involvement of inhibitory effects on CSSC uptake (DeHart et al., 2018). Knockdown of either TRX-domain protein 12 (TXNDC12) (Yu, Zhu, Wang, & Jiang, 2023) or GLRX5 (Lee, You, Shin, & Roh, 2020) have furthermore been shown to sensitize cells to ferroptosis as triggered by erastin, showing how the different cellular redox systems are functionally linked and that the exact molecular mechanisms of erastin leading to ferroptosis are yet not fully understood. In this section, we aim to specifically scrutinize a few aspects of erastin mechanisms of action that warrant further studies.

2.1.1 On- and off-target activities of erastin

With VDACs and xCT having been implicated as targets of erastin, as discussed above, one should consider that the compound, as for all small molecules, might also have other protein targets in cells. In recent non-biased proteome analyses performed with crude cell lysates, erastin was found to clearly affect 3061 proteins, while another ferroptosis-inducing compound ML210 impacted 2828 proteins, and 2550 proteins were altered upon treatment with BSO; among these proteins as many as 2278 were altered in all three experiments (Kudryashova et al., 2023). Such data suggest that proteome changes during ferroptosis are far-reaching, and perhaps surprisingly reproducible, but the actual protein targets of the compounds used in these cases to trigger the cell death are not identified through such analyses. A more direct method evaluates shifts in thermostability of specific proteins upon treatment of cells with a specific compound, assuming that when a compound binds a specific protein the stability of that protein becomes altered. Another method to study target engagement employs alkyne-derivatives and click chemistry to directly link a compound in question to its plausible protein targets. Using these approaches to detect potential protein targets of erastin, a study recently found more than 800 potential erastin targets in cells, of which five were identified with higher confidence and further validated in subsequent experiments. These five additional erastin targets were identified as monoacylglycerol lipase ABHD6, the epoxide hydrolase 1 EPHX1, the mitochondrial-processing peptidase subunit α-PMPCA, the puromycin-sensitive aminopeptidase NPEPPS, and the saccharopine dehydrogenase-like oxidoreductase SCCPDH (Li, Liu, Wang, & Wang, 2024). It remains unknown how targeting any of these proteins, in addition to VDACs and xCT, might contribute to the cellular or pharmacological effects of erastin.

2.1.2 Sulfasalazine—similar results as erastin?

Sulfasalazine (SSZ) is, citing its PubChem entry "an azobenzene consisting of diphenyldiazene having a carboxy substituent at the 4-position, a hydroxy substituent at the 3-position and a 2-pyridylaminosulphonyl substituent at the 4′-position, known to act as a non-steroidal anti-inflammatory drug, an antiinfective agent, a gastrointestinal drug, a glutathione transferase inhibitor, a drug allergen, and a ferroptosis inducer" (PubChem ID: 5339). SSZ has been used since the 1950s as an anti-inflammatory drug and is still proscribed for use in treatment of ulcerative colitis and rheumatoid arthritis (https://www.drugs.com/sulfasalazine.html). Although well tolerated in animals and patients, in cell cultures, the death-triggering effects of SSZ were suggested, similarly to erastin, to be due to direct inhibition of xCT and thus inhibition of the cellular uptake of cystine (Ogihara et al., 2019). Also like erastin, it was shown that iron plays a role in SSZ-triggered cell death, hence pointing to ferroptosis (Liu et al., 2022). One study suggested that these effects could, in turn, be modulated by the expression level of the estrogen receptor (Yu et al., 2019). Early proteomics analyses using 2-D gel electrophoresis combined with mass spectrometry, showed that SSZ affected the levels of hundreds of proteins in treated cells, with a profile compatible with its anti-inflammatory properties (Endo et al., 2014). In more recent proteome analyses, SSZ was again shown to have wide-ranging effects and also triggered significant redox-related changes in many proteins; interestingly in the opposite direction as occurred with treatment of the cells with H_2O_2 in an attempt to directly trigger oxidative stress (Sun et al., 2019). In the context of the present review, we here wish to conclude that both erastin and SSZ can trigger ferroptosis coincident with their inhibition of xCT, but the cellular effects of both compounds are wide-ranging. We suggest that a mere inhibition of Cys uptake is unlikely to be the sole mechanism of action of these compounds.

2.1.3 Impacts of erastin or SSZ on cellular cystine uptake and Cys or GSH levels

Both erastin and SSZ are usually suggested to induce ferroptosis through inhibition of cystine uptake and, thus, by compromising cellular Cys availability. What is the evidence for this? One important original study showing these effects reported that erastin, sulfasalazine and the kinase inhibitor sorafenib, all triggered typical ferroptotic cell death; this was also coupled with blockage of cystine uptake and Glu release. Furthermore, similar effects on Glu release were induced by silencing the SLC7A11 component of xCT with siRNAs, but not by silencing the large-neutral amino acid transporter SLC7A5 (LAT1) (Dixon et al., 2014). What could the arguments be that other mechanisms of action would be involved? For one, whole-body SLC7A11 knockout mice are viable (Chen et al., 2023; Hamashima et al., 2017; Zhang et al., 2022b), suggesting that merely inhibition of xCT cannot likely be sufficient to trigger ferroptosis in any critical cell types in the body, at least in normal conditions. Second, cytotoxicity of sulfasalazine was found to depend upon ASCT2-dependent glutamine (Gln) uptake and Glu dehydrogenase (GLUD)-mediated α-ketoglutarate (α-KG) production, suggesting that effects on glutamino-lysis and not on Cys uptake are involved (Okazaki et al., 2019). Third, could perhaps an additional reason for erastin- and sulfasalazine-mediated inhibition of Cys uptake be that these compounds trigger NRF2 activation, driving *de novo* Cys biosynthesis by transsulfuration (see above), thus blocking uptake of cystine from the extracellular space by competitive inhibition from interacellularly synthesized cysteine? We propose that this is theoretically possible, although having no definite evidence for it to be the case. We merely conclude that it might be wise not to consider erastin or sulfasalazine as "pure" or "single-acting" inhibitors of xCT, as is unfortunately often done in current ferroptosis literature.

2.2 Protection/prevention against ferroptosis by iron-chelators

With the definition of ferroptosis being a type of cell death blocked by iron chelators (see above), it becomes a truism that iron chelators prevent ferroptosis. But what are the mechanisms of action for the protective effects of iron chelators? A likely explanation should be that chelation of free iron cations prevents their reactivity with H_2O_2 through the well-known Fenton reaction producing the very reactive hydroxyl radical (Merkofer, Kissner, Hider, Brunk, & Koppenol, 2006). If occurring close to a cell membrane, this could produce the lipid peroxides seen in ferroptosis (Gutteridge, 1986), although the exact molecular species involved in the iron-promoted lipid peroxidation may be more complex (Minotti & Aust, 1989). It should be noted that, chemically, copper ions can partake in the same type of peroxidation reactions as iron and, indeed, the term "cuproptosis" has also been coined for the copper-triggered cell death that would otherwise likely be highly analogous to ferroptosis (Chen, Min, & Wang, 2022). In this context, it should be emphasized that cell death triggered by excessive oxidative stress has long been known, described, and discussed (Orrenius, Gogvadze, & Zhivotovsky, 2007; Sies, Berndt, & Jones, 2017; Sies, 2020; Zhivotovsky & Nicotera, 2020). Thus, both ferroptosis and cuproptosis are likely to be forms of cell death triggered by oxidative stress that specifically involve Fenton-chemistry and lipid peroxidation.

2.3 Association with lipid peroxidation

Lipids, especially polyunsaturated fatty acids including phospholipids and other species, can easily be further oxidized with several enzymatic pathways leading to their oxidation through lipoxygensases, such as during metabolism of arachidonic acid as part of the synthesis of leukotrienes, prostaglandins or thromboxane species (Kuhn, Banthiya, & van Leyen, 2015; Radmark, Werz, Steinhilber, & Samuelsson, 2015; Ricciotti & FitzGerald, 2011). Lipids may also be oxidized in a non-regulated manner through the Fenton reaction-like free-radical mechanisms mentioned above. It has furthermore been proposed that 15-lipoxygenases might specifically produce lipid peroxides as a mechanism to trigger enzymatically regulated initiation of ferroptosis (Bayir et al., 2020; Stoyanovsky et al., 2019). Irrespective of the possible distinctions between "regulated" or "non-regulated" iron-triggered lipid peroxidation, it seems clear that the adverse accumulation of lipid peroxides, presumably mainly in cellular membranes, is the actual event leading to the type of cell death that defines ferroptosis. Using mass spectrometry approaches this has been clearly shown, mainly demonstrating accumulation of phospholipid hydroperoxide species during ferroptosis (Sparvero et al., 2021; Wiernicki et al., 2020), perhaps with specific sites, such as the mitochondrial membrane, being more important than other (Lyamzaev, Panteleeva, Simonyan, Avetisyan, & Chernyak, 2023).

3. Role of GPX4 in ferroptosis

The notion that GPX4 can protect cells from cell death through ferroptosis due to its capacity to reduce lipid peroxides and thereby prevent their accumulation, is widely acknowledged. Still, there are several unanswered questions regarding the exact mechanisms of the links between GPX4 and ferroptosis, as was also previously discussed at detail by Matilde Maiorino, Marcus Conrad and Fulvio Ursini (Maiorino, Conrad, & Ursini, 2018; Ursini & Maiorino, 2020). It is important to note that GPX4 is not the only suppressor of ferroptosis in cells that may act through reduction, directly or indirectly, of lipid peroxides, or prevent their accumulation (see Section 1.6, below). This may help to explain why or how the impact of GPX4 seems to be important in relation to ferroptosis in only certain cell types or growth conditions, and not in others. Here we shall specifically point out a few questions regarding the roles of GPX4 in relation to ferroptosis where we see a clear need for further studies.

3.1 Kinetics and substrate specificities of GPX4

GPX4 was originally discovered in 1982 by Ursini and coworkers as an enzyme reducing phosphatidylcholine hydroperoxide using GSH and in 1982 they showed that the enzyme was a selenoprotein (Ursini, Maiorino, & Gregolin, 1985). Later it was shown that GPX4 specifically associates with phospholipids in membranes to facilitate the reduction of phospholipid hydroperoxides, and that cardiolipin is a preferred binding partner

(Cozza et al., 2017). Interestingly, an Arginine-to-Histidine mutation at residue 152 (R152H), found in cases of Sedaghatian-type Spondylometaphyseal Dysplasia (SSMD) that encompasses severe neurodevelopmental dysfunction in human, mainly leads to dissociation of GPX4 from cardiolipin rather than affecting its catalytic turnover, as such (Roveri et al., 2023). Comparisons of the substrate specificities and enzyme activities of recombinant forms of GPX1, GPX2 and GPX4 side-by-side under identical conditions recently confirmed that GPX4 is the only of these three enzymes that can reduce phophatidylcholine hydroperoxides; GPX1 is far more efficient than the other GPX isoenzymes in reduction of other peroxides, including H_2O_2, cumene hydroperoxide, tertbutyl hydroperoxide or other fatty acid hydroperoxides including those derived from linoleic acid, arachidonic acid or eicosapentaenoic acid (Schwarz et al., 2023). Thus, GPX4 seems to be particularly and uniquely specialized for reduction of membrane-associated phospholipid hydroperoxides, which underlies its important role in preventing the accumulation of toxic levels of these molecular species in cells.

3.2 Results that question the exact mechanisms of GPX4 targeting in ferroptosis

RSL3 (PubChem ID: 1750826) was discovered by the Stockwell group in 2008 in a screen for compounds being specifically toxic towards tumor cells harboring mutated small GTPases, and was thus named RSL3 for "RAS-selective lethal" (Yang & Stockwell, 2008). It was later used together with erastin in the study published in 2012 naming ferroptosis (Dixon et al., 2012). In 2014, the group identified GPX4 as the target of RSL3 using chemoproteomics with an RSL3-derivative for identification of cellular protein targets; that study also showed that erastin treatment caused cellular GSH depletion (Yang et al., 2014). These findings, together with the cellular effects of additional ferroptosis inducers, suggested that GPX4 was the target of RSL3 and thus a key suppressor of ferroptosis. The study also showed that RAS-transformed cells were more sensitive to GSH depletion using BSO than control cells, that overexpression or knockdown of GPX4 prevented or accentuated ferroptosis, respectively, and that this cell death was associated with increased lipid peroxidation (Yang et al., 2014). It was thus a surprise when the group of Ursini and coworkers could not find any inhibition of purified GPX4 using RSL3, while if they included cell lysate and/or the 14-3-3-3 protein under reducing conditions, inhibition could be detected (Vuckovic et al., 2020).

Our own group (ESJA) confirmed Ursini's results using recombinant GPX4 selenoprotein by showing that RSL3 is not a direct inhibitor of the enzyme, but we were even more surprised when we discovered that RSL3 is, instead, a direct inhibitor of TrxR1 both in recombinant form and in a cellular context (Cheff et al., 2023). A similar result was found for another ferroptosis inducing compound, ML162 (Cheff et al., 2023). Noteworthy, while both RSL3 and ML162 triggered typical ferroptosis in cells when used at very low nM-range concentrations; another more selective inhibitor of TrxR1, TRi-1, did not (Sabatier et al., 2021; Stafford et al., 2018). RSL3, ML162 and ML210 (the latter of which could not directly inhibit either GPX4 or TrxR1) all triggered a cell death blocked by ferrostatin or iron chelators, while the cell death triggered by TRi-1 at low µM range was not affected by ferrostatin or iron chelators. When ferrostatin was used to protect cells treated with RSL3, the cytotoxicity of RSL3 was blunted and instead induced cell death at higher concentrations, very closely matching that obtained with TRi-1 (Cheff et al., 2023). Interestingly, we also observed that RSL3, ML162 and ML210 all affected the migration of GPX4 in SDS-PAGE, suggestive of downstream indirect effects on GPX4 when the ferroptosis inducing compounds were used in cells (Cheff et al., 2023).

Very recently the group of Marcus Conrad addressed the question regarding the possible RSL3-GPX4 interaction using HEK293 cells transfected with Streptavidin-tagged GPX4, to enable affinity purification of the enzyme with or without treatment with ferroptosis inducing compounds (Nakamura et al., 2024). They confirmed the altered GPX4 migration in SDS-PAGE after treatment with RSL3, and they show inhibition of the enzyme activity of the purified enzyme from RSL3-treated cells or upon treatment in vitro. However, they used incubation of cells with 10 µM RSL3 (1000-fold higher concentrations than those triggering ferroptosis) or incubation of purified enzyme using $0-1$ µM RSL3 in vitro at 37 °C for 30 min, resulting in approximately 60% inhibition of GPX4 activity during an additional 30 min incubation, as judged from the quantitation of remaining phosphatidylcholine peroxide (Nakamura et al., 2024). We judge those results to display a rather weak inhibition of GPX4 by RSL3. At this stage, we wish to conclude that it seems clear that ferroptosis inducing compounds, including RSL3, ML162 and ML210, somehow affect cellular GPX4 and its migration in subsequent SDS-PAGE analyses, but that could be due to indirect mechanisms. The Conrad group posits that mammalian cell-specific posttranslational modifications of GPX4 might be required for RSL3-mediated inhibition of

GPX4, which would explain why the protein affinity purified from mammalian cells is inhibited by RSL3 whereas the *E. coli*-expressed recombinant selenoprotein is not (Cheff et al., 2023). Clearly additional studies will be needed to resolve whether RSL3 interacts directly with GPX4 in vivo, whether that is dependent on mammalian posttranslational modifications, and indeed the exact molecular mechanisms of action by which RSL3 induces ferroptosis.

Only a few selenoprotein-encoding complete genetic knockout models cause embryonic lethality in mice, including *Txnrd1* encoding cytosolic TrxR1 (Bondareva et al., 2007; Jakupoglu et al., 2005), *Txnrd2* encoding mitochondrial TrxR2 (Conrad et al., 2004), and *Gpx4* encoding both the cytosolic, nuclear and mitochondrial forms of GPX4 (Seiler et al., 2008; Yant et al., 2003). A similar phenotype is seen with expression of mutant GPX4 variants with the catalytic Sec residue mutated to either inactive Ala or Ser (Ingold et al., 2015), clearly showing that it is a lack of GPX4 catalytic activity that leads to embryonic lethality. Interestingly, targeted deletion of only the nuclear form of GPX4 yields no overt phenotype and the only effect that could be discovered was that of increased sperm abnormalities (Conrad et al., 2005), which likely relates to the fact that GPX4 is also a moonlighting protein converted during spermatogenesis into a structural component of sperm (Ursini et al., 1999). When mitochondrial GPX4 was deleted, this gave a similar effect with a lack of overt phenotype but abnormal sperm maturation, thus suggesting that the cytosolic form of GPX4 is the essential form (Schneider et al., 2009). Nonetheless it remained unclear whether lack of cytosolic GPX4 was embryonically lethal due to increased ferroptosis, or whether other mechanisms lay behind the lethality upon its deletion. It could be reminiscent of the genetic deletion of *Txnrd1*, which yielded early embryonic lethality without signs of tissue damage due to increased oxidative stress and, instead, exhibited a lack of formation of mesoderm indicating distorted intracellular differentiation or signaling programs (Bondareva et al., 2007; Dagnell, Schmidt, & Arner, 2018).

Further addressing the question of the reasons for lethality upon GPX4 deletions, Marcus Conrad and coworkers replaced GPX4 in a mouse model with a Sec-to-Cys "knock-in" variant of the enzyme (*Gpx4^{cys}* allele), which displays a much lower turnover with hydroperoxide substrates yet is not totally inactive. With an inbred C57Bl/6J mouse strain background, crossing parents heterozygous for this knock-in (*Gpx4^{sec/cys}*), homozygous Sec-Cys mice (*Gpx4^{cys/cys}*), unlike mice homozygous for either the null-mutation ($Gpx4^{-/-}$) or a totally catalytically inactive Sec-Ser knock-in ($Gpx4^{ser/ser}$), which each exhibit 100% embryonic lethality by embryonic-day 7.5 (E7.5) (Ingold et al., 2015), died not until after organogenesis (E11.5-12.5) (Ingold et al., 2018). This finding was further studied with a Tamoxifen-induced conditional allele-switch strategy in adult mice, wherein most cells in the mice switched from being heterozygous across a functionally wildtype floxed allele, ($Gpx4^{fl}$) (Friedmann Angeli et al., 2014) for either the Sec-to-Cys or the Sec-to-Ser allele ($Gpx4^{fl/cys}$ or $Gpx4^{fl/ser}$, respectively), to being hemizygous for only the Sec-to-Cys or the Sec-to-Ser allele ($Gpx4^{-/cys}$ or $Gpx4^{-/ser}$, respectively). That strategy showed that the induced adult-onset $Gpx4^{-/ser}$ mice, like induced adult-onset $Gpx4^{-/-}$ mice, died within 11 days of Tamoxifen-induced allelic conversion; both showing similar renal pathology. By contrast, the induced adult-onset $Gpx4^{-/cys}$ mice thrived for at least 40 d after Tamoxifen-induced allelic conversion (Friedmann Angeli et al., 2014; Ingold et al., 2018). Moreover, cultured $Gpx4^{cys/cys}$ cells displayed typical ferroptotic cell death when challenged, suggesting that Sec-containing GPX4 indeed protects cells from ferroptosis (Ingold et al., 2018). Notably, the cells overexpressing the Sec-to-Cys variant of GPX4 could also survive genetic co-deletion of the tRNA for selenocysteine, and thus survive deletion of all other selenoproteins (Ingold et al., 2018).

Of note, when the $Gpx4^{cys}$ allele was studied on a hybrid strain-background 129S6SvEV x C57Bl/6J) all $Gpx4^{cys/cys}$ pups survived to birth; however they failed to thrive postnatally and, by postnatal day 18, all had developed severe seizures and needed to be sacrificed (Ingold et al., 2018). Necropsies revealed these pups lacked PV+ interneurons (Ingold et al., 2018). Considering the available data, it can thus be concluded that cytosolic GPX4 is clearly an essential enzyme, unique in being able to reduce phospholipid peroxide species, and that the enzyme typically protects cells in culture from ferroptosis. It has not yet been unequivocally shown that the embryonic lethality of inbred $Gpx4^{cys/cys}$ fetuses, nor the lack of PV+ interneurons in the hybrid-strain $Gpx4^{cys/cys}$ pups, was due to ferroptosis, as assessing either would be extremely technically difficult. However, in the latter case, an increased number of TUNEL-positive cells were seen in the cortex, associated with increased astrogliosis and neuroinflammation (Ingold et al., 2018). This phenotype is consistent with cell death by diverse mechanisms, including apoptosis or ferroptosis, thereafter leading to inflammatory responses.

Interestingly, recent reports suggest that GPX4 knockdown does not induce ferroptosis via a mechanism that might involve failure to reduce lipid peroxides, but rather induces overaccumulation of ferrous iron, leading to ferroptosis (Wei et al., 2022).

4. FSP1 and other enzymes affecting ferroptosis in parallel with GPX4

With ferroptosis being defined as a cell death triggered by iron-dependent lipid peroxidation, this suggests that all proteins or enzymes that may limit either iron reactivity or accumulation of lipid peroxides may prevent ferroptosis. Indeed, several such systems have been found, but shall here only briefly be introduced with this review article mainly being focused on the links between cysteine homeostasis and ferroptosis. These additional enzyme systems acting in parallel with GPX4 are, in contrast to GPX4, typically not considered to be dependent upon GSH.

4.1 FSP1

In 2019, both the groups of Marcus Conrad and of James Olzmann reported that the flavoprotein apoptosis-inducing factor mitochondrial 2 (AIFM2) efficiently protected cells from ferroptosis upon deletion of GPX4, thereby renaming AIFM2 to Ferroptosis suppressor protein 1 (FSP1). They also showed that FSP1 expression levels closely correlate with resistance of cells to ferroptosis, that inhibition of FSP1 sensitizes cells to ferroptosis, and that myristoylated membrane-associated FSP1 acts by reducing ubiquinone coenzyme Q10 (CoQ10) in a reaction using NADPH, with reduced ubiquinol being able to trap lipid peroxyl radicals and thus suppress propagation of lipid peroxidation (Bersuker et al., 2019; Doll et al., 2019). The Conrad group furthermore showed that FSP1 can alternatively use Vitamin K in catalytic lipid peroxide trapping (Mishima et al., 2022). Recently the same group also reported that a number of 3-phenylquinazolinone compounds can target FSP1, trigger its release from the membrane, and lead to aggregation of the enzyme and intracellular phase separation, also correlating with increased sensitivity to ferroptosis (Nakamura et al., 2023). The structural features of FSP1 were solved and the catalytic cycle described (Lv et al., 2023). FSP1, like many other antioxidant enzymes, is also an NRF2 target, and is upregulated in many cancer cells (Bersuker et al., 2019; Koppula et al., 2022; Muller et al., 2023). Thus, pharmacological inhibition of FSP1 makes cancer cells and tumors more prone to cell death, thereby showing that FSP1 holds promise as an anticancer therapy target (Bersuker et al., 2019; Doll et al., 2019; Hendricks et al., 2023; Muller et al., 2023; Nakamura et al., 2023).

4.2 DHODH

In 2021 it was reported by the group of Boyi Gan that dihydroorotate dehydrogenase (DHODH) could protect cells from ferroptosis by regenerating ubiquinol, independently from FSP1 (Mao et al., 2021). This notion was subsequently challenged by the group of Marcus Conrad, showing that the study had used inhibitors of DHODH at high concentrations, which also inhibited FSP1 (Mishima et al., 2023). This word of caution was replied to by Boyi Gan et al., arguing that not only the use of inhibitors showed that DHODH plays a role in protection of cells against ferroptosis, but also genetic evidence (Mao, Liu, Yan, Olszewski, & Gan, 2023). This topic remains unresolved, but perhaps it can be wise at this stage to remain open to the possibility that DHODH can indeed protect certain cells against ferroptosis, depending upon overall cellular context, metabolic state and expression of other proteins and enzymes affecting iron-dependent peroxidation events and thus steady-state levels of phospholipid hydroperoxides.

4.3 MBOAT1 and MBOAT2

Very recently it was shown that membrane-bound O-acyltransferases MBOAT1 and MBOAT2, being transcriptional targets upregulated by the estrogen and androgen receptors, respectively, can "remodel" cellular phospholipids and thus make cells more resistant to ferroptosis in a GPX4- as well as FSP1-indendent manner (Liang et al., 2023). The remodeling reaction catalyzed by the MBOAT enzymes involves acyl transfer reactions, and the enzymes accept several different phospholipids as their substrates, including phosphatidylcholine (Gijon, Riekhof, Zarini, Murphy, & Voelker, 2008), the peroxide of which is a unique GPX4 substrate (see above). There are more than 10 family members of MBOATs in human (Masumoto et al., 2015) and thus the evaluation of their individual possible importance in relation to ferroptosis will require significant additional studies in the forthcoming years.

4.4 Proteins affecting iron status

If ferroptosis is defined as being iron-dependent, then iron homeostasis should likely affect the extent of ferroptotic cell death. Indeed, many studies have reported such observations. This includes NRF2-driven HERC2 and VAMP8 indirectly increasing ferritin levels, thus shown to block ferroptosis (Anandhan et al., 2023); transferrin-knockout mice being more prone to ferroptosis in the liver (Yu et al., 2020); alpha-Enolase 1 (ENO1) moonlighting as an mRNA-binding protein suppressing iron regulatory protein 1 (IRP1) expression leading to inhibition of mitoferrin-1 (Mfrn1) expression and subsequent repression of mitochondrial iron-induced ferroptosis (Zhang et al., 2022a); or prominin-2 leading to export from cells of iron and thereby leading to suppression of ferroptosis (Brown et al., 2019). Additional examples include NUPR1-mediated LCN2 expression that blocks ferroptosis by preventing iron accumulation and subsequent oxidative damage (Liu et al., 2021a), or loss of ferritin through autophagy leading to increased ferroptosis (Hou et al., 2016). There are many examples in the literature showing that distorted iron status can be linked to modulation of ferroptosis.

5. Genetic and pharmacologic models that question the exact roles of cystine, Cys, xCT, GSH, or GSH synthesis in ferroptosis

The discovery of ferroptosis using compounds that disrupt xCT activity and the long-association of ferroptosis with xCT inhibition, has led to the prevalent working hypothesis wherein diminished CSSC is the first step on a chain-reaction that diminishes cellular Cys, cellular GSH, and GSH-mediated transit of reducing power to GPX4 (Dixon et al., 2012; Jiang et al., 2021; Yan et al., 2021). However, as detailed above, aspects of this inhibitor- and cell culture-based relationship have proven recalcitrant to subsequent validations, in particular using other inhibitors, genetic approaches, or whole-animal models. Moreover, the inhibitors that formed the basis of this hypothesis are now known to have often poorly characterized mechanisms of action on xCT, and to have wide-ranging and incompletely defined off-target effects. Similar concerns are now being raised about next-generation ferroptosis-inducing compounds targeting GPX4. In the best-case scenario, the working hypothesis is conditionally correct, but requires other predisposing conditions to result in ferroptosis. As an example of this, one can look to the original discovery of erastin, which reported that it only reliably caused non-apoptotic cell death in cell lines that carried mutant HRAS and the SV40 small-T antigen (Dolma et al., 2003). Clearly there are many missing pieces to the puzzle regarding what exact conditions that will trigger ferroptosis in a living cell. In this section we will critically discuss the components of the hypothesized chain-reaction from extracellular CSSC to lipid peroxidation and cell death, to better clarify which are critical components; which are, perhaps, only "passenger components" arising from the ferroptosis-inducing treatments, yet not playing a causal role; and which are perhaps "conditionally causal", for example reliably inducing ferroptosis in cell culture conditions but not necessarily relevant in vivo.

5.1 CSSC and xCT

The hypothesis that restricting access of cells to CSSC can induce ferroptosis, whether these are HRAS-mutant/small-T antigen expressing cancer cells or not, is problematic. As overviewed in Section 1.1.2, above, although CSSC is typically the sole source of Cys in cell culture media and, in these conditions, knockdown of xCT can promote ferroptosis (Xu, Huang, Zhu, & Guo, 2022), it is unclear whether CSSC is reliably available to cells in vivo. Moreover, it is very clear that xCT is not essential in vivo, as discussed above. Since there is not a mechanism for splanchnic tissues to export CSSC into the blood plasma (Fig. 2), CSSC in vivo appears to be predominantly supplied by the gut upon digestion of food-protein. Thus, if CSSC restriction either extracellularly or by blockage of xCT induced ferroptosis, then intermittent fasting should induce ferroptosis in vivo, at least in mutant HRAS/small-T antigen expressing cancers, which has not been reported. Moreover, the MGI database of The Jackson Laboratories (https://www. informatics.jax.org/marker/MGI:1347355) reports that full-body homozygous xCT (SLC7A11)-null mice have mild pigmentation defects but are overtly healthy, consistent with CSSC being a convenient source of Cys for cells to exploit when available, e.g., after a protein-rich meal, but not a necessary source in any known cell type in vivo. Clearly other sources of Cys can supplant CSSC uptake in vivo. This obvious conclusion, nonetheless, largely annuls approaches targeting xCT to induce ferroptosis in vivo. Additionally, for the model of CSSC/xCT/Cys targeting to induce ferroptosis via lowering GSH levels and, thereby, lowering GSH-mediated trafficking of reducing power to GPX4, these treatments must have impacts on GSH greater than other treatments that do not induce ferroptosis. However, as also discussed above, cells and organisms, including cancer cells in organisms or in patients, are highly tolerant of treatments, notably BSO administration, that precipitously lower GSH levels. We shall in the next section further discuss this aspect in specific relation to ferroptosis.

5.2 GSSG/GSH and GGT

As overviewed in Sections 1.1 and 1.2 above, GSH and GSSG are likely to be a more important source of systemic Cys than CSSC, arising from intermediary metabolism in splanchnic tissues. Of these, GSH is more abundant in plasma but, until a GSH exporter can be identified, this appears to arise from secreted GSSG at the expense of reducing power captured from secreted protein-thiols (see above). However, mice homozygous-null for a knockout of GGT1, the predominant GGT enzyme (with a second enzyme, GGT5, and phenotypes associated with its disruption, appearing to be restricted to immune cells (Shi, Han, Habib, Matzuk, & Lieberman, 2001), show a mild growth retardation phenotype (Lieberman et al., 1996). It is noteworthy that full-body pharmacologic disruption of GSH biosynthesis in mice with BSO dramatically lowers plasma GSSG+GSH levels, with this being well tolerated and, although BSO treatment compromises growth of some cancer cell lines in culture (Wei et al., 2022), systemic BSO treatment has not been useful in either killing or diminishing growth in tumor-bearing mice (Coshan-Gauthier & Kirkpatrick, 1989; Zhu, Hu, Wu, & Hu, 2014). The ability of BSO to sensitize tumors to other therapies has varied (Geroni et al., 1993; Soble & Dorr, 1988; Tsutsui et al., 1986), and it is unclear which of these effects arise from restricting GSH as a source of Cys versus via restricting its other activities, e.g., in drug metabolism pathways or its other activities (Tanaka, Kurokawa, Matsuno, Matsumoto, & Hayashida, 2008).

5.3 GCL disruptions

The target enzyme of BSO is the Glu-Cys ligase enzyme (GCL), which is a heterodimer composed of a catalytic subunit, GCLC, that is responsible for the chemical reaction, and a modulatory subunit, GCLM, that confers product (GSH)-inhibition on the holoenzyme (Chen, Shertzer, Schneider, Nebert, & Dalton, 2005). Genetic models targeting GCL have not supported the working hypothesis of ferroptosis induction (see Section 2, above). Thus, whereas GCLM-null mouse models have exhibited diverse and usually modest alterations in GSH levels, they have not caused ferroptosis (Botta et al., 2008; Franklin et al., 2009; Nakamura et al., 2011). More recently, using conditional disruption of GCLC in whole mice, the team of Isaak Harris has shown that many cells in whole mouse models tolerate disruption of GCLC. Whereas that result is consistent with decades of research using BSO to inhibit GCL activity, as an inhibitor, the specificities and potential off-target activities of BSO could never be fully assessed, whereas by disrupting the gene encoding GCLC, the new studies more precisely hone-in on the specific activities of GCL; in this way, they have shown that T cell-specific genetic disruption of GCLC and therefore, of GSH biosynthesis, does not cause cell death but, instead, compromises the functions of regulatory T cells (Kurniawan et al., 2020). Additional insights to the in vivo requirements for GSH biosynthesis are anticipated from the continuation of this work.

5.4 GSR KO

Another tenet in the prevalent working hypothesis that targeting xCT induces ferroptosis by indirectly interfering with GSH-mediated transit of reducing power to GPX4, is that other treatments that interfere with GSH-mediated transit of reducing power to GPX4 should also induce ferroptosis. TrxR1 cannot reduce GSSG (Arnér, 2009b), meaning that only one primary disulfide reductase in the cytosol, GSR, can directly reduce GSSG → 2 GSH (Miller & Schmidt, 2019). Nonetheless, TRX1 can also reduce GSSG (Gromer, Merkle, Schirmer, & Becker, 2002). Importantly, similar to the situation wherein we long recognized that there must be a mechanism to transport GSH into mitochondria (Booty et al., 2015) even though the transporter was only recently identified (Liu et al., 2023) (see Section 1.2.3, above), we know that cells can sustain the highly reduced intracellular status of the GSH+GSSG pool in the absence of GSR (Rogers et al., 2004). Although this has been attributed to "cross-trafficking" of reducing power from the TRX system to the GSH system (Zhang, Du, Zhang, Lu, & Holmgren, 2014), the ability of TrxR1/GSR- and TRX1/TrxR1/GSR-null mouse livers to also sustain the reduced status of the GSH+GSSG pool indicates that completely TrxR1-independent mechanisms should also be at play here (Miller & Schmidt, 2020). To date, the only such TrxR1- and GSR-independent mechanisms that have been implicated involve *de novo* synthesis of Cys and cellular export of GSSG (Sections 1.1 and 1.2, above) which, although still untested, we predict should be less efficient at trafficking disulfide reducing power to target enzymes including GPX4.

6. Potential roles of the thioredoxin system in ferroptosis

With TrxR1 being a highly important selenoprotein for support of reductive power in cells (Arnér & Holmgren, 2000; Arnér, 2009a; Miller et al., 2018; Nordberg & Arnér, 2001) that was suggested together with

GPX4 to be one of only two key essential selenoproteins in human cells (Santesmasses & Gladyshev, 2022), and considering that the ferroptosis inducing compounds RSL3 and ML162 efficiently inhibit TrxR1 (Cheff et al., 2023), it could have been hypothesized that TrxR1 would be another key enzyme for protection of cells against ferroptosis. However, there's no clear data at this moment suggesting this would be the case. When TrxR1 is targeted with inhibitors, cancer cells typically die, but not with ferroptosis features (Cheff et al., 2023; Sabatier et al., 2021; Stafford et al., 2018), and although the enzyme is essential for embryonic development (Bondareva et al., 2007; Jakupoglu et al., 2005) it can be conditionally genetically deleted from adult tissues without typical signs of ferroptosis induction (Dagnell et al., 2018; Eriksson et al., 2015; Iverson et al., 2013; Miller et al., 2018; Prigge et al., 2012; Prigge et al., 2017; Rollins et al., 2010). Importantly, the cell death triggered by TrxR1 inhibitors is not prevented by ferrostatin or iron chelators, in contrast to that seen with low (but not higher) concentrations of RSL3 and ML162 (Cheff et al., 2023). Of special note here, for the in vivo functions of TrxR1, is that mouse livers living under TrxR1-deficient conditions have not been found to undergo precipitous ferroptosis nor, remarkably, are disulfide reductase-deficient malignant hepatocellular carcinomas induced in such mouse livers via chemical carcinogenesis (McLoughlin et al., 2019). Based on those robust, living, non-ferroptotic mouse systems, we consider it formally possible, yet nonetheless unlikely, that diminishing import of CSSC via xCT could have a large enough impact on the trafficking of electrons to GPX4 to induce ferroptosis. More likely perhaps, we suggest that alternative mechanisms, in particular ones that do not require CSSC, Cys, or GSH restriction, need to be considered for how cell death can be triggered in normal cells upon TrxR1 targeting, such as during embryogenesis, if at all, such as with the lack of overt phenotypes upon conditional knockout in adult tissues.

We consider that TrxR1 does not seem to be a *bona fide* suppressor of ferroptosis but should rather be viewed of more as an important pillar of antioxidant defense systems and reductive pathways of cells, in general. In cancer cells, however, it may be another case, with several groups having shown that cancer cells are sensitive to TrxR1 targeting, both in cell culture and in vivo (Chew, Lu, Bradshaw, & Holmgren, 2008; Gencheva & Arner, 2022; Lu, Chew, & Holmgren, 2007; Mandal et al., 2010; Scalcon, Bindoli, & Rigobello, 2018; Stafford et al., 2018; Urig & Becker, 2006; Wang et al., 2012; Zhang et al., 2016). The mechanisms of cell death also in these cases do not, however, seem to be typical for ferroptosis.

Whereas NADPH-GSR is a likely primary source of reducing power for sustaining the highly reduced status of the GSH:GSSG pool in cells, NADPH-TrxR1 appears to play the predominant role is supplying the reducing power to convert nutritional CSSC entering the cytosol, regardless of from xCT, GGT, or cystinosin, into 2 Cys (Fig. 3). Thus, biochemical studies showed that NADPH-TrxR1/TRP14 is at least 5-fold more efficient at catalyzing reduction of CSSC into 2 Cys than is NADPH-TrxR1/TRX1 and, unlike other redoxins, TRP14 activity with CSSC is not competed by substrates such as protein-disulfides (Pader et al., 2014). If Cys availability becomes limiting for cytosolic Cys production, whether from pathways involving CSSC→xCT→CSSC, GSSG→GGT→CSSC, or proteolysis→cystinosin→CSSC, as often posited for ferroptosis induction, then the redoxin systems, in particular NADPH-TrxR1/TRP14, should be particularly critical for preventing ferroptosis. This prediction, however, has at least not yet been borne out experimentally.

NADPH-TrxR1 is the primary reductase reducing cytosolic TRX1 or TRP14, although a cytosolic isoform of GLRX2, entitled GLRX2c, can also reduce the active site disulfide of TRX1 and, thereby, might serve as a conduit for cross-trafficking reducing power between the TrxR1 and GSH pathways (Zhang et al., 2014). However, decades of treatment of cells, animals, and patients with the potent TrxR1-inhibiting drug AFN have shown that this inhibition does not cause detectable ferroptosis in normal cells (Arnér, 2009b), nor has it yet been shown effective at treating cancers (Arner & Holmgren, 2006), probably suggesting that ferroptosis is not induced by AFN. Another more specific TrxR1 inhibitor, TRi-1, similarly has not been reported to induce ferroptosis (Sabatier et al., 2021; Stafford et al., 2018), nor are other known TrxR1 inhibitors considered to be ferroptosis inducing compounds (Gencheva & Arner, 2022). Also, mouse and cell models with genetic disruptions of TrxR1 have been developed, yet none to our knowledge, including those in cancers, have been reported to cause ferroptosis (Bondareva et al., 2007; Jakupoglu et al., 2005; Mandal et al., 2010; McLoughlin et al., 2019; Peng et al., 2016). Possibly the lack of ferroptosis in all these observations could be explained if the NADPH-GSR system, either via GLRX-mediated reduction of CSSC or via trafficking of reducing power via GLRX2c→→TRP14, complemented the effects of loss of TrxR1 on CSSC reduction. However, subsequent development of TR/GR-null mouse liver, cell, or cancer models, clearly shows that, even in the absence of any primary NADPH-driven cytosolic disulfide

Table 1 Presumed players in ferroptosis and outstanding questions regarding their exact roles.

Player	Expected role	Outstanding Questions with regards to ferroptosis	Sections
CSSC	Source of Cys	Cells survive without CSSC. CSSC not likely limiting source of Cys in vivo	1.1, 5.1
Cys	Limit GSH synthesis	Required by all cells. Unclear that cell death by Cys-restriction is normally ferroptosis	1.1
GSH	Limit GPX4 activity	Most cells tolerate near-total GSH depletion	1.2
GSH:GSSG	Limit GPX4 activity	Regulated by GSR and GSSG-export. Not clear this ratio can increase to incapacitate GPX4	1.2, 5.2
H2O2	Precursor for HO.-	Few sources of H_2O_2 associated with ferroptosis	2.1.2, 2.2
xCT	Import of CSSC	xCT-null mice survive. CSSC not only source of Cys in vivo	5.1
TRP14	Reduce CSSC–>2Cys	TRP14-null mice overtly normal. TRP14-disruption does not cause ferroptosis	1.5
TRX1	Reduce CSSC–>2Cys	TRX1-null livers survive. TRX1-disruption does not cause ferroptosis	1.5
TrxR1	Fuels TRP14, TRX1	TrxR1-livers and cells survive. TrxR1-disruption does not cause ferroptosis	1.5
GCL	Synthesis of GSH	GCLM-null cells and mice survive. At least some GCLC-null T cell types survive	1.2, 5.3
GSR	Reduce GSSG–>2GSH	GSR-null mice overtly normal. GSR-disruption does not cause ferroptosis	1.5, 5.4
Erastin	Inhibit xCT	Causes ferroptosis, off-target activities. Mechanisms unclear, possibly indirect	1.4.1

SSZ	Inhibit xCT	Off-target activities. Mechanisms of xCT disruption unclear and possibly indirect	2.1
AFN	Inhibit TrxR1	Whole-body AFN tolerated. Inhibits TrxR1 but does not cause ferroptosis	1.5, 6
Tri-1	Inhibit TrxR1	Diverse cells tolerate Tri-1. Inhibits TrxR1 but does not cause ferroptosis	1.5, 3.2, 6
BSO	Inhibit GCL	Depletes GSH in cells and mice, but tolerated and rarely associated with ferroptosis	1.2, 2.1.1
RSL3	Inhibit GPX4	Causes ferroptosis, mechanisms unclear. Inhibits TrxR1; unclear if it inhibits GPX4 directly	3.2, 6
ML162	Inhibit GPX4	Causes ferroptosis, mechanisms unclear. Inhibits TrxR1; unclear if it inhibits GPX4 directly	3.2, 6
Death	non-apoptotic	Death mechanisms? Unclear whether mitochondrial leakage and apoptosis are involved	2, 5

This table summarizes some of the outstanding questions regarding Cys metabolism and its relation to ferroptosis discussed in this review. References are given to the sections of this review for further details. See also Fig. 5.

reductase system, this does not result in an inability to reduce CSSC thus causing detectable ferroptosis (Eriksson et al., 2015; McLoughlin et al., 2019; Prigge et al., 2017).

The TrxR1 inhibitor, AFN, and the TrxR1- and TR/GR-null models discussed in the previous paragraph all have the characteristic of inducing strong chronic activation of the Nrf2 system, which plausibly could protect cells against ferroptosis (Cebula, Schmidt, & Arner, 2015; Iverson et al., 2013; Locy et al., 2012; Prigge et al., 2017; Suvorova et al., 2009). However, more recently we have developed fibroblast- and liver-specific mouse models lacking TRX1 (Prigge et al., 2017), and cell culture- or full body mouse-models lacking TRP14 (Doka et al., 2016; Doka et al., 2020), none of which show measurable activation of Nrf2 (McLoughlin et al., 2019;

Prigge et al., 2017), yet these also have not resulted in ferroptosis. Moreover, even TRX1/TR/GR-null mouse liver or liver cancer models (McLoughlin et al., 2019; Prigge et al., 2017), or TRP14/TRX1/TR/GR-null mouse livers (EES, unpublished) display overt evidence of ferroptosis. As such, since the thioredoxin system is crucial for CSSC reduction, we must seriously question what role, if any, CSSC reduction plays in triggering ferroptosis.

Based on these robust, living, non-ferroptotic mouse systems, both cancerous and non-cancerous, we consider it formally possible, yet nonetheless unlikely, that diminishing import of CSSC via xCT could have a large enough impact on the trafficking of electrons to GPX4 to induce ferroptosis. Instead, we suggest that alternative mechanisms should be considered for how ferroptosis is induced, in particular ones that do not require CSSC, Cys, or GSH restriction. Several outstanding questions regarding individual chemicals, enzymes, and pharmaceutical compounds, as discussed above, concerning their roles in ferroptosis mechanisms, are summarized in Table 1.

7. Concluding remarks

Ferroptosis is an interesting and provocative topic. In the context of cancer, it is also a potentially targetable physiological response that has been intensely studied for over a decade, but remains mechanistically mysterious. This remarkable mode of cell death was identified as early as 2012, yet targeting ferroptosis-susceptibility to treat cancer remains an elusive goal. Much in the literature, unfortunately, presents ferroptosis as a rather straightforward mode of cell death, with xCT-mediated CSSC import and subsequent potency of the GSH system as a critical limiting factor in supporting GPX4 activity, and with iron-propelled lipid peroxidation leading to cell death prevented by either GPX4 or FSP1. In this review we have tried to emphasize that ferroptosis indeed likely derives from oxidative stress-related damage to cell membranes, yet that that the current rather strict definition or view of ferroptosis conceptually limits the interpretations and understanding of underlying mechanisms of action. Several of the small molecule compounds currently used in studies of ferroptosis are likely not as specific in their target engagements as is often presented. Several suspected mechanistic components, including CSSC, xCT, Cys, and GSH, are potentially "false-flags", arising from models and conditions that do not reflect cancers in patients. Critically, an oversimplified view of ferroptosis can perhaps distract focus away from its true mechanisms of action. We have here attempted to bring attention to experimental systems usually not used in ferroptosis research, and the physiology of Cys metabolism, surprisingly often not considered in mechanistic models of ferroptosis, aiming to integrate these topics. We hope that our review will stimulate discussions, kindly asking the many colleagues in the field of ferroptosis research to be conservative in interpretations of the observed effects, and being open minded with regards to the complexity of biological systems.

Acknowledgments

We wish to thank all our group members and collaborators for the many past, ongoing and future efforts in the exciting field of redox biology, undoubtedly with great promise for novel discoveries of major importance in biology and medicine. We also wish to thank our funding agencies for support of the work in our own groups. E.S.J.A. wishes to acknowledge funding from Karolinska Institutet, The Knut and Alice Wallenberg Foundations KAW 2019.0059, The Swedish Cancer Society 21 1463 Pj, The Swedish Research Council 2021-02214, The Cayman Biomedical Research Institute (CABRI), National Laboratories Excellence program under the National Tumor Biology Laboratory project 2022-2.1.1-NL-2022-00010 and the Hungarian Thematic Excellence Programme TKP2021-EGA-44 and The National Research, Development and Innovation Office NKFIH grant ED_18-1-2019-0025. E.E.S. acknowledges funding from the United States National Institutes of Health grants AG040020, AG055022, OD026444, DK123738, and P30GM140964 as well as from the Hungarian Eötvös Loránd Kutatási Hálózat Foundation ELKH, grant #15002, the Hungarian Magyar Tudományos Akadémia MTA, grant K-22 #143769, and a Distinguished Guest Fellowship from the Hungarian MTA #AT02023-26.

References

Anandhan, A., Dodson, M., Shakya, A., Chen, J., Liu, P., Wei, Y., ... Zhang, D. D. (2023). *Sci Adv, 9*, eade9585.

Anderson, M. E., & Meister, A. (1980). *The Journal of Biological Chemistry, 255*, 9530–9533.

Arner, E. S., & Holmgren, A. (2000). *European Journal of Biochemistry/FEBS, 267*, 6102–6109.

Arner, E. S., & Holmgren, A. (2006). *Seminars in Cancer Biology, 16*, 420–426.

Arnér, E. S. J. (2009a). *Biochimica et Biophysica Acta, 1790*, 495–526.

Arnér, E. S. J. (2009b). *Biochimica et Biophysica Acta, 1790*, 495–526.

Ballatori, N., Krance, S. M., Marchan, R., & Hammond, C. L. (2009). *Molecular Aspects of Medicine, 30*, 13–28.

Bayir, H., Anthonymuthu, T. S., Tyurina, Y. Y., Patel, S. J., Amoscato, A. A., Lamade, A. M., ... Kagan, V. E. (2020). *Cell Chem Biol, 27*, 387–408.

Bersuker, K., Hendricks, J. M., Li, Z., Magtanong, L., Ford, B., Tang, P. H., ... Olzmann, J. A. (2019). *Nature, 575*, 688–692.

Bondareva, A. A., Capecchi, M. R., Iverson, S. V., Li, Y., Lopez, N. I., Lucas, O., ... Schmidt, E. E. (2007). *Free Radical Biology & Medicine, 43*, 911–923.

Bonifacio, V. D. B., Pereira, S. A., Serpa, J., & Vicente, J. B. (2021). *British Journal of Cancer, 124*, 862–879.

Booty, L. M., King, M. S., Thangaratnarajah, C., Majd, H., James, A. M., Kunji, E. R., & Murphy, M. P. (2015). *FEBS Letters, 589*, 621–628.

Botta, D., White, C. C., Vliet-Gregg, P., Mohar, I., Shi, S., McGrath, M. B., ... Kavanagh, T. J. (2008). *Drug Metabolism Reviews, 40*, 465–477.

Brigelius-Flohe, R. (2006). *Biological Chemistry, 387*, 1329–1335.

Brigelius-Flohe, R., & Maiorino, M. (2013). *Biochimica et Biophysica Acta, 1830*, 3289–3303.

Brown, C. W., Amante, J. J., Chhoy, P., Elaimy, A. L., Liu, H., Zhu, L. J., ... Mercurio, A. M. (2019). *Developmental Cell, 51*, 575–586.e4.

Cassier-Chauvat, C., Marceau, F., Farci, S., Ouchane, S., & Chauvat, F. (2023). *Antioxidants (Basel), 12*.

Cebula, M., Schmidt, E. E., & Arner, E. S. (2015). *Antioxidants & Redox Signaling, 23*, 823–853.

Cheff, D. M., Huang, C., Scholzen, K. C., Gencheva, R., Ronzetti, M. H., Cheng, Q., ... Arner, E. S. J. (2023). *Redox Biol, 62*, 102703.

Chen, Y., Lu, T., Liu, Y., Liu, Y., Bai, S., Chen, Q., ... Wu, X. (2023). *In Vitro Cellular & Developmental Biology. Animal, 59*, 729–737.

Chen, L., Min, J., & Wang, F. (2022). *Signal Transduct Target Ther, 7*, 378.

Chen, Y., Shertzer, H. G., Schneider, S. N., Nebert, D. W., & Dalton, T. P. (2005). *The Journal of Biological Chemistry, 280*, 33766–33774.

Chew, E. H., Lu, J., Bradshaw, T. D., & Holmgren, A. (2008). *The FASEB Journal, 22*, 2072–2083.

Chiang, F. F., Chao, T. H., Huang, S. C., Cheng, C. H., Tseng, Y. Y., & Huang, Y. C. (2022). *Int J Mol Sci, 23*.

Cole, S. P., & Deeley, R. G. (2006). *Trends in Pharmacological Sciences, 27*, 438–446.

Combs, J. A., & DeNicola, G. M. (2019). *Cancers (Basel), 11*.

Conrad, M., Jakupoglu, C., Moreno, S. G., Lippl, S., Banjac, A., Schneider, M., ... Brielmeier, M. (2004). *Molecular and Cellular Biology, 24*, 9414–9423.

Conrad, M., Moreno, S. G., Sinowatz, F., Ursini, F., Kolle, S., Roveri, A., ... Bornkamm, G. W. (2005). *Molecular and Cellular Biology, 25*, 7637–7644.

Coshan-Gauthier, R. K., & Kirkpatrick, D. L. (1989). *Experimental Cell Biology, 57*, 273–280.

Cozza, G., Rossetto, M., Bosello-Travain, V., Maiorino, M., Roveri, A., Toppo, S., ... Ursini, F. (2017). *Free Radical Biology & Medicine, 112*, 1–11.

Dagnell, M., Pace, P. E., Cheng, Q., Frijhoff, J., Ostman, A., Arner, E. S. J., ... Winterbourn, C. C. (2017). *The Journal of Biological Chemistry, 292*, 14371–14380.

Dagnell, M., Schmidt, E. E., & Arner, E. S. J. (2018). *Free Radical Biology & Medicine, 115*, 484–496.

Daher, B., Vucetic, M., & Pouyssegur, J. (2020). *Front Oncol, 10*, 723.

DeHart, D. N., Fang, D., Heslop, K., Li, L., Lemasters, J. J., & Maldonado, E. N. (2018). *Biochemical Pharmacology, 148*, 155–162.

Deneke, S. M., & Fanburg, B. L. (1989). *The American Journal of Physiology, 257*, L163–L173.

Dixon, S. J., Lemberg, K. M., Lamprecht, M. R., Skouta, R., Zaitsev, E. M., Gleason, C. E., ... Stockwell, B. R. (2012). *Cell, 149*, 1060–1072.

Dixon, S. J., Patel, D. N., Welsch, M., Skouta, R., Lee, E. D., Hayano, M., ... Stockwell, B. R. (2014). *Elife, 3*, e02523.

Doka, E., Ida, T., Dagnell, M., Abiko, Y., Luong, N. C., Balog, N., ... Nagy, P. (2020). *Sci Adv, 6*, eaax8358.

Doka, E., Pader, I., Biro, A., Johansson, K., Cheng, Q., Ballago, K., ... Nagy, P. (2016). *Sci Adv, 2*, e1500968.

Doll, S., Freitas, F. P., Shah, R., Aldrovandi, M., Da Silva, M. C., Ingold, I., ... Conrad, M. (2019). *Nature, 575*, 693–698.

Dolma, S., Lessnick, S. L., Hahn, W. C., & Stockwell, B. R. (2003). *Cancer Cell, 3*, 285–296.

Endo, W., Arito, M., Sato, T., Kurokawa, M. S., Omoteyama, K., Iizuka, N., ... Kato, T. (2014). *Modern Rheumatology/the Japan Rheumatism Association, 24*, 844–850.

Eriksson, S., Prigge, J. R., Talago, E. A., Arner, E. S., & Schmidt, E. E. (2015). *Nat Commun, 6*, 6479.

Finkelstein, J. D. (1998). *European Journal of Pediatrics, 157*(Suppl 2), S40–S44.

Flohe, L., Toppo, S., Cozza, G., & Ursini, F. (2011). *Antioxidants & Redox Signaling, 15*, 763–780.

Fomenko, D. E., Marino, S. M., & Gladyshev, V. N. (2008). *Molecules and Cells, 26*, 228–235.

Franklin, C. C., Backos, D. S., Mohar, I., White, C. C., Forman, H. J., & Kavanagh, T. J. (2009). *Molecular Aspects of Medicine, 30*, 86–98.

Friedmann Angeli, J. P., Schneider, M., Proneth, B., Tyurina, Y. Y., Tyurin, V. A., Hammond, V. J., ... Conrad, M. (2014). *Nature Cell Biology, 16*, 1180–1191.

Gencheva, R., & Arner, E. S. J. (2022). *Annual Review of Pharmacology and Toxicology, 62*, 177–196.

Geroni, C., Pesenti, E., Tagliabue, G., Ballinari, D., Mongelli, N., Broggini, M., ... Grandi, M. (1993). *International Journal of Cancer. Journal International du Cancer, 53*, 308–314.

Gijon, M. A., Riekhof, W. R., Zarini, S., Murphy, R. C., & Voelker, D. R. (2008). *The Journal of Biological Chemistry, 283*, 30235–30245.

Green, J. A., Vistica, D. T., Young, R. C., Hamilton, T. C., Rogan, A. M., & Ozols, R. F. (1984). *Cancer Research, 44*, 5427–5431.

Griffith, O. W., Bridges, R. J., & Meister, A. (1981). *Proc Natl Acad Sci U S A, 78*, 2777–2781.

Griffith, O. W., & Meister, A. (1979a). *Proc Natl Acad Sci U S A, 76*, 5606–5610.

Griffith, O. W., & Meister, A. (1979b). *The Journal of Biological Chemistry, 254*, 7558–7560.

Gromer, S., Merkle, H., Schirmer, R. H., & Becker, K. (2002). *Methods in Enzymology, 347*, 382–394.

Guo, X., Schmiege, P., Assafa, T. E., Wang, R., Xu, Y., Donnelly, L., ... Li, X. (2022). *Cell, 185*, 3739–3752.e18.

Gutteridge, J. M. (1986). *FEBS Letters, 201*, 291–295.

Hamashima, S., Homma, T., Kobayashi, S., Ishii, N., Kurahashi, T., Watanabe, R., ... Fujii, J. (2017). *Free Radical Research, 51*, 851–860.

Hansen, R. E., Roth, D., & Winther, J. R. (2009). *Proc Natl Acad Sci U S A, 106*, 422–427.

Harris, I. S., Treloar, A. E., Inoue, S., Sasaki, M., Gorrini, C., Lee, K. C., ... Mak, T. W. (2015). *Cancer Cell, 27*, 211–222.

Hendricks, J. M., Doubravsky, C. E., Wehri, E., Li, Z., Roberts, M. A., Deol, K. K., ... Olzmann, J. A. (2023). *Cell Chem Biol, 30*, 1090–1103.e7.

Holmgren, A. (2000). *Antioxidants & Redox Signaling, 2*, 811–820.

Hou, W., Xie, Y., Song, X., Sun, X., Lotze, M. T., Zeh, H. J., 3rd, ... Tang, D. (2016). *Autophagy, 12*, 1425–1428.

Ingold, I., Aichler, M., Yefremova, E., Roveri, A., Buday, K., Doll, S., ... Conrad, M. (2015). *The Journal of Biological Chemistry, 290*, 14668–14678.

Ingold, I., Berndt, C., Schmitt, S., Doll, S., Poschmann, G., Buday, K., ... Conrad, M. (2018). *Cell, 172*, 409–422.e21.

Israel, B. A., Jiang, L., Gannon, S. A., & Thorpe, C. (2014). *Free Radical Biology & Medicine, 69*, 129–135.

Iverson, S. V., Eriksson, S., Xu, J., Prigge, J. R., Talago, E. A., Meade, T. A., ... Schmidt, E. E. (2013). *Free Radical Biology & Medicine, 63*, 369–380.

Jakupoglu, C., Przemeck, G. K., Schneider, M., Moreno, S. G., Mayr, N., Hatzopoulos, A. K., ... Conrad, M. (2005). *Molecular and Cellular Biology, 25*, 1980–1988.

Jiang, X., Stockwell, B. R., & Conrad, M. (2021). *Nature Reviews. Molecular Cell Biology, 22*, 266–282.

Jones, D. P., Carlson, J. L., Mody, V. C., Jr, Cai, J., Lynn, M. J., & Sternberg, P., Jr (2000). *Free Radical Biology and Medicine, 28*, 625–635.

Jyotsana, N., Ta, K. T., & DelGiorno, K. E. (2022). *Front Oncol, 12*, 858462.

Kabil, O., Vitvitsky, V., Xie, P., & Banerjee, R. (2011). *Antioxidants & Redox Signaling, 15*, 363–372.

Kaya, A., Lee, B. C., & Gladyshev, V. N. (2015). *Antioxidants & Redox Signaling, 23*, 814–822.

Kelner, M. J., & Montoya, M. A. (2000). *Biochemical and Biophysical Research Communications, 269*, 366–368.

Kim, H. Y., & Gladyshev, V. N. (2007). *The Biochemical Journal, 407*, 321–329.

Kobayashi, S., Ikeda, Y., Shigeno, Y., Konno, H., & Fujii, J. (2020). *Amino Acids, 52*, 555–566.

Kojer, K., Bien, M., Gangel, H., Morgan, B., Dick, T. P., & Riemer, J. (2012). *The EMBO Journal, 31*, 3169–3182.

Koppula, P., Lei, G., Zhang, Y., Yan, Y., Mao, C., Kondiparthi, L., ... Gan, B. (2022). *Nat Commun, 13*, 2206.

Kudryashova, O. M., Nesterenko, A. M., Korzhenevskii, D. A., Sulyagin, V. K., Tereshchuk, V. M., Belousov, V. V., & Shokhina, A. G. (2023). *Data, 8*, 119.

Kuhn, H., Banthiya, S., & van Leyen, K. (2015). *Biochimica et Biophysica Acta, 1851*, 308–330.

Kurniawan, H., Franchina, D. G., Guerra, L., Bonetti, L., Baguet, L. S., Grusdat, M., ... Brenner, D. (2020). *Cell Metabolism, 31*, 920–936.e7.

Lee, J., & Roh, J. L. (2022). *Antioxidants (Basel), 11*.

Lee, J., You, J. H., Shin, D., & Roh, J. L. (2020). *Theranostics, 10*, 7775–7786.

Le, D. T., Liang, X., Fomenko, D. E., Raza, A. S., Chong, C. K., Carlson, B. A., ... Gladyshev, V. N. (2008). *Biochemistry*.

Liang, D., Feng, Y., Zandkarimi, F., Wang, H., Zhang, Z., Kim, J., ... Jiang, X. (2023). *Cell, 186*, 2748–2764.e22.

Lieberman, M. W., Wiseman, A. L., Shi, Z. Z., Carter, B. Z., Barrios, R., Ou, C. N., ... Matzuk, M. M. (1996). *Proc Natl Acad Sci U S A, 93*, 7923–7926.

Liu, L. J., Jiang, Z., Wang, P., Qin, Y. L., Xu, W., Wang, Y., ... Jiang, C. Y. (2021b). *Front Microbiol, 12*, 768283.

Liu, J., Lou, C., Zhen, C., Wang, Y., Shang, P., & Lv, H. (2022). *Metallomics, 14*.

Liu, J., Song, X., Kuang, F., Zhang, Q., Xie, Y., Kang, R., ... Tang, D. (2021a). *Nat Commun, 12*, 647.

Liu, Y., Liu, S., Tomar, A., Yen, F. S., Unlu, G., Ropek, N., ... Birsoy, K. (2023). *Science, 382*, 820–828.

Li, Z., Liu, D., Wang, Y., & Wang, C. (2024). *Chemical Research in Toxicology, 37*, 109–116.

Locy, M. L., Rogers, L. K., Prigge, J. R., Schmidt, E. E., Arner, E. S., & Tipple, T. E. (2012). *Antioxidants & Redox Signaling, 17*, 1407–1416.

Lu, J., Chew, E. H., & Holmgren, A. (2007). *Proc Natl Acad Sci U S A, 104*, 12288–12293.

Lu, S. C. (2013). *Biochimica et Biophysica Acta, 1830*, 3143–3153.

Lv, Y., Liang, C., Sun, Q., Zhu, J., Xu, H., Li, X., ... Zhu, D. (2023). *Nat Commun, 14*, 5933.

Lyamzaev, K. G., Panteleeva, A. A., Simonyan, R. A., Avetisyan, A. V., & Chernyak, B. V. (2023). *Biophys Rev, 15*, 875–885.

Mahagita, C., Grassl, S. M., Piyachaturawat, P., & Ballatori, N. (2007). *American Journal of Physiology. Gastrointestinal and Liver Physiology, 293*, G271–G278.

Maiorino, M., Conrad, M., & Ursini, F. (2018). *Antioxidants & Redox Signaling, 29*, 61–74.

Mandal, P. K., Schneider, M., Kolle, P., Kuhlencordt, P., Forster, H., Beck, H., ... Conrad, M. (2010). *Cancer Research, 70*, 9505–9514.

Marti-Andres, P., Finamor, I., Torres, I., Perez, S., Rius-Perez, S., Colino-Lage, H., ... Sastre, J. (2024). *EMBO J.* In press. PMID: 38811853.

Mao, C., Liu, X., Yan, Y., Olszewski, K., & Gan, B. (2023). *Nature, 619*, E19–E23.

Mao, C., Liu, X., Zhang, Y., Lei, G., Yan, Y., Lee, H., ... Gan, B. (2021). *Nature, 593*, 586–590.

Masumoto, N., Lanyon-Hogg, T., Rodgers, U. R., Konitsiotis, A. D., Magee, A. I., & Tate, E. W. (2015). *Biochemical Society Transactions, 43*, 246–252.

McLoughlin, M. R., Orlicky, D. J., Prigge, J. R., Krishna, P., Talago, E. A., Cavigli, I. R., ... Schmidt, E. E. (2019). *Proc Natl Acad Sci U S A, 116*, 11408–11417.

Meira, W., Daher, B., Parks, S. K., Cormerais, Y., Durivault, J., Tambutte, E., ... Vucetic, M. (2021). *Cancers (Basel), 13*.

Meister, A., Griffith, O. W., Novogrodsky, A., & Tate, S. S. (1979). *Ciba Foundation Symposium*, 135–161.

Merkofer, M., Kissner, R., Hider, R. C., Brunk, U. T., & Koppenol, W. H. (2006). *Chemical Research in Toxicology, 19*, 1263–1269.

Miller, C. G., Holmgren, A., Arner, E. S. J., & Schmidt, E. E. (2018). *Free Radical Biology & Medicine, 127*, 248–261.

Miller, C. G., & Schmidt, E. E. (2019). *British Journal of Pharmacology, 176*, 532–543.

Miller, C. G., & Schmidt, E. E. (2020). *Antioxidants & Redox Signaling, 33*, 1158–1173.

Minotti, G., & Aust, S. D. (1989). *Chemico-Biological Interactions, 71*, 1–19.

Miseta, A., & Csutora, P. (2000). *Molecular Biology and Evolution, 17*, 1232–1239.

Mishima, E., Ito, J., Wu, Z., Nakamura, T., Wahida, A., Doll, S., ... Conrad, M. (2022). *Nature, 608*, 778–783.

Mishima, E., Nakamura, T., Zheng, J., Zhang, W., Mourao, A. S. D., Sennhenn, P., & Conrad, M. (2023). *Nature, 619*, E9–E18.

Morgan, B., Ezerina, D., Amoako, T. N., Riemer, J., Seedorf, M., & Dick, T. P. (2013). *Nature Chemical Biology, 9*, 119–125.

Morgan, B., Sobotta, M. C., & Dick, T. P. (2011). *Free Radical Biology & Medicine, 51*, 1943–1951.

Muller, F., Lim, J. K. M., Bebber, C. M., Seidel, E., Tishina, S., Dahlhaus, A., ... von Karstedt, S. (2023). *Cell Death and Differentiation, 30*, 442–456.

Nagy, P., Karton, A., Betz, A., Peskin, A. V., Pace, P., O'Reilly, R. J., ... Winterbourn, C. C. (2011). *The Journal of Biological Chemistry, 286*, 18048–18055.

Nakamura, B. N., Fielder, T. J., Hoang, Y. D., Lim, J., McConnachie, L. A., Kavanagh, T. J., & Luderer, U. (2011). *Endocrinology, 152*, 2806–2815.

Nakamura, T., Hipp, C., Santos Dias Mourao, A., Borggrafe, J., Aldrovandi, M., Henkelmann, B., ... Conrad, M. (2023). *Nature, 619*, 371–377.

Nakamura, T., Ito, J., Mourao, A. S. D., Wahida, A., Nakagawa, K., Mishima, E., & Conrad, M. (2024). *Cell Rep Methods*, 100710.

Netto, L. E. S., & Machado, L. (2022). *The FEBS Journal, 289*, 5480–5504.

Nordberg, J., & Arner, E. S. (2001). *Free Radical Biology & Medicine, 31*, 1287–1312.

Oestreicher, J., & Morgan, B. (2019). *Biochemistry and Cell Biology = Biochimie et Biologie Cellulaire, 97*, 270–289.

Ogihara, K., Kikuchi, E., Okazaki, S., Hagiwara, M., Takeda, T., Matsumoto, K., ... Oya, M. (2019). *Cancer Science, 110*, 1431–1441.

Okazaki, S., Umene, K., Yamasaki, J., Suina, K., Otsuki, Y., Yoshikawa, M., ... Nagano, O. (2019). *Cancer Science, 110*, 3453–3463.

Orrenius, S., Gogvadze, V., & Zhivotovsky, B. (2007). *Annual Review of Pharmacology and Toxicology, 47*, 143–183.

Pader, I., Sengupta, R., Cebula, M., Xu, J., Lundberg, J. O., Holmgren, A., ... Arner, E. S. (2014). *Proc Natl Acad Sci U S A, 111*, 6964–6969.

Peng, X., Gimenez-Cassina, A., Petrus, P., Conrad, M., Ryden, M., & Arner, E. S. (2016). *Sci Rep, 6*, 28080.

Prigge, J. R., Coppo, L., Martin, S. S., Ogata, F., Miller, C. G., Bruschwein, M. D., ... Schmidt, E. E. (2017). *Cell Rep, 19*, 2771–2781.

Prigge, J. R., Eriksson, S., Iverson, S. V., Meade, T. A., Capecchi, M. R., Arnér, E. S. J., & Schmidt, E. E. (2012). *Free Radical Biology & Medicine, 52*, 803–810.

Radmark, O., Werz, O., Steinhilber, D., & Samuelsson, B. (2015). *Biochimica et Biophysica Acta, 1851*, 331–339.

Requejo, R., Hurd, T. R., Costa, N. J., & Murphy, M. P. (2010). *The FEBS Journal, 277*, 1465–1480.

Reznik, N., & Fass, D. (2022). *FEBS Letters, 596*, 2859–2872.

Ricciotti, E., & FitzGerald, G. A. (2011). *Arteriosclerosis, Thrombosis, and Vascular Biology, 31*, 986–1000.

Rogers, L. K., Tamura, T., Rogers, B. J., Welty, S. E., & Hansen, T. N. (2004). and Smith, C. V. *Toxicological Sciences: an Official Journal of the Society of Toxicology, 82*, 367–373.

Rollins, M. F., van der Heide, D. M., Weisend, C. M., Kundert, J. A., Comstock, K. M., Suvorova, E. S., ... Schmidt, E. E. (2010). *Journal of Cell Science, 123*, 2402–2412.

Rosell, R., Jain, A., Codony-Servat, J., Jantus-Lewintre, E., Morrison, B., Ginesta, J. B., & Gonzalez-Cao, M. (2023). *Cancer Biol Med, 20*, 500–518.

Roveri, A., Di Giacinto, F., Rossetto, M., Cozza, G., Cheng, Q., Miotto, G., ... Ursini, F. (2023). *Redox Biol, 64*, 102806.

Russel, M., Model, P., & Holmgren, A. (1990). *Journal of Bacteriology, 172*, 1923–1929.

Sabatier, P., Beusch, C. M., Gencheva, R., Cheng, Q., Zubarev, R., & Arner, E. S. J. (2021). *Redox Biol, 48*, 102184.

Santesmasses, D., & Gladyshev, V. N. (2022). *Biomolecules, 12*.

Sato, M., Kusumi, R., Hamashima, S., Kobayashi, S., Sasaki, S., Komiyama, Y., ... Sato, H. (2018). *Sci Rep, 8*, 968.

Scalcon, V., Bindoli, A., & Rigobello, M. P. (2018). *Free Radical Biology & Medicine, 127*, 62–79.

Scalise, M., Pochini, L., Console, L., Losso, M. A., & Indiveri, C. (2018). *Front Cell Dev Biol, 6*, 96.

Schneider, M., Forster, H., Boersma, A., Seiler, A., Wehnes, H., Sinowatz, F., ... Conrad, M. (2009). *The FASEB Journal, 23*, 3233–3242.

Schwarz, M., Loser, A., Cheng, Q., Wichmann-Costaganna, M., Schadel, P., Werz, O., ... Kipp, A. P. (2023). *Redox Biol, 59*, 102593.

Seiler, A., Schneider, M., Forster, H., Roth, S., Wirth, E. K., Culmsee, C., ... Conrad, M. (2008). *Cell Metabolism, 8*, 237–248.

Shi, Z. Z., Han, B., Habib, G. M., Matzuk, M. M., & Lieberman, M. W. (2001). *Molecular and Cellular Biology, 21*, 5389–5395.

Sies, H. (2020). *Antioxidants (Basel), 9*.

Sies, H., Berndt, C., & Jones, D. P. (2017). *Annual Review of Biochemistry, 86*, 715–748.

Soble, M. J., & Dorr, R. T. (1988). *Anticancer Research, 8*, 17–22.

Sparvero, L. J., Tian, H., Amoscato, A. A., Sun, W. Y., Anthonymuthu, T. S., Tyurina, Y. Y., ... Bayir, H. (2021). *Angewandte Chemie (International Ed. in English), 60*, 11784–11788.

Stafford, W. C., Peng, X., Olofsson, M. H., Zhang, X., Luci, D. K., Lu, L., ... Arner, E. S. J. (2018). *Sci Transl Med, 10*.

Stenersen, J., Kobro, S., Bjerke, M., & Arend, U. (1987). *Comp Biochem Physiol C Comp Pharmacol Toxicol, 86*, 73–82.

Stipanuk, M. H., Dominy, J. E., Jr., Lee, J. I., & Coloso, R. M. (2006). *The Journal of Nutrition, 136*, 1652S–1659S.

Stipanuk, M. H., & Ueki, I. (2011). *Journal of Inherited Metabolic Disease, 34*, 17–32.

Stockwell, B. R., & Jiang, X. (2020). *Cell Chem Biol, 27*, 365–375.

Stoyanovsky, D. A., Tyurina, Y. Y., Shrivastava, I., Bahar, I., Tyurin, V. A., Protchenko, O., ... Kagan, V. E. (2019). *Free Radical Biology & Medicine, 133*, 153–161.

Sun, W., Dai, L., Yu, H., Puspita, B., Zhao, T., Li, F., ... Nordlund, P. (2019). *Redox Biol, 24*, 101168.

Suvorova, E. S., Lucas, O., Weisend, C. M., Rollins, M. F., Merrill, G. F., Capecchi, M. R., & Schmidt, E. E. (2009). *PLoS One, 4*, e6158.

Talwar, D., Miller, C. G., Grossmann, J., Szyrwiel, L., Schwecke, T., Demichev, V., ... Dick, T. P. (2023). *Nat Metab, 5*, 660–676.

Tanaka, T., Kurokawa, H., Matsuno, K., Matsumoto, S., & Hayashida, Y. (2008). *Anticancer Research, 28*, 2663–2668.

Tate, S. S., & Meister, A. (1981). *Molecular and Cellular Biochemistry, 39*, 357–368.

Thuillier, A., Ngadin, A. A., Thion, C., Billard, P., Jacquot, J. P., Gelhaye, E., & Morel, M. (2011). *Int J Evol Biol, 2011*, 938308.

Toppo, S., Flohe, L., Ursini, F., Vanin, S., & Maiorino, M. (2009). *Biochimica et Biophysica Acta, 1790*, 1486–1500.

Trachootham, D., Alexandre, J., & Huang, P. (2009). *Nature Reviews. Drug Discovery, 8*, 579–591.

Trommelen, J., Tomé, D., & van Loon, L. J. (2021). *Clinical Nutrition Open Science, 36*, 43–55.

Tsutsui, K., Komuro, C., Ono, K., Nishidai, T., Shibamoto, Y., Takahashi, M., & Abe, M. (1986). *International Journal of Radiation Oncology, Biology, Physics, 12*, 1183–1186.

Urig, S., & Becker, K. (2006). *Seminars in Cancer Biology, 16*, 452–465.

Ursini, F., Heim, S., Kiess, M., Maiorino, M., Roveri, A., Wissing, J., & Flohe, L. (1999). *Science (New York, N.Y.), 285*, 1393–1396.

Ursini, F., & Maiorino, M. (2020). *Free Radical Biology & Medicine, 152*, 175–185.

Ursini, F., Maiorino, M., & Gregolin, C. (1985). *Biochimica et Biophysica Acta, 839*, 62–70.

Van Der Kolk, D. M., Vellenga, E., Muller, M., & de Vries, E. G. (1999). *Advances in Experimental Medicine and Biology, 457*, 187–198.

Vuckovic, A. M., Bosello Travain, V., Bordin, L., Cozza, G., Miotto, G., Rossetto, M., ... Roveri, A. (2020). *FEBS Letters, 594*, 611–624.

Wang, L., Yang, Z., Fu, J., Yin, H., Xiong, K., Tan, Q., ... Zeng, H. (2012). *Free Radical Biology & Medicine, 52*, 898–908.

Wei, S., Yu, Z., Shi, R., An, L., Zhang, Q., Zhang, Q., ... Wang, H. (2022). *BMC Cancer, 22*, 881.

Wiernicki, B., Dubois, H., Tyurina, Y. Y., Hassannia, B., Bayir, H., Kagan, V. E., ... Vanden Berghe, T. (2020). *Cell Death Dis, 11*, 922.

Winterbourn, C. C. (2013). *Methods in Enzymology, 528*, 3–25.

Winterbourn, C. C., & Hampton, M. B. (2008). *Free Radical Biology & Medicine, 45*, 549–561.

Winterbourn, C. C., & Metodiewa, D. (1999). *Free Radical Biology & Medicine, 27*, 322–328.

Wray, J. L., Campbell, E. I., Roberts, M. A., & Gutierrez-Marcos, J. F. (1998). *Chemico-Biological Interactions, 109*, 153–167.

Xu, T., Ding, W., Ji, X., Ao, X., Liu, Y., Yu, W., & Wang, J. (2019). *Journal of Cellular and Molecular Medicine, 23*, 4900–4912.

Xu, R., Huang, Y., Zhu, D., & Guo, J. (2022). *Free Radical Biology & Medicine, 184*, 158–169.

Yang, W. S., SriRamaratnam, R., Welsch, M. E., Shimada, K., Skouta, R., Viswanathan, V. S., ... Stockwell, B. R. (2014). *Cell, 156*, 317–331.

Yang, W. S., & Stockwell, B. R. (2008). *Chemistry & Biology, 15*, 234–245.

Yant, L. J., Ran, Q., Rao, L., Van Remmen, H., Shibatani, T., Belter, J. G., ... Prolla, T. A. (2003). *Free Radical Biology & Medicine, 34*, 496–502.

Yan, H. F., Zou, T., Tuo, Q. Z., Xu, S., Li, H., Belaidi, A. A., & Lei, P. (2021). *Signal Transduct Target Ther, 6*, 49.

Yu, Y., Jiang, L., Wang, H., Shen, Z., Cheng, Q., Zhang, P., ... Wang, F. (2020). *Blood, 136*, 726–739.

Yu, H., Yang, C., Jian, L., Guo, S., Chen, R., Li, K., ... Liu, S. (2019). *Oncology Reports, 42*, 826–838.

Yu, H., Zhu, K., Wang, M., & Jiang, X. (2023). *Clin Transl Sci, 16*, 1957–1971.

Zhang, H., Du, Y., Zhang, X., Lu, J., & Holmgren, A. (2014). *Antioxidants & Redox Signaling, 21*, 669–681.

Zhang, T., Sun, L., Hao, Y., Suo, C., Shen, S., Wei, H., ... Zhang, H. (2022a). *Nat Cancer, 3*, 75–89.

Zhang, B., Zhang, J., Peng, S., Liu, R., Li, X., Hou, Y., ... Fang, J. (2016). *Expert Opin Ther Pat,* 1–10.

Zhang, X., Zheng, C., Gao, Z., Chen, H., Li, K., Wang, L., ... Meng, Y. (2022b). *Cardiovascular Drugs and Therapy/Sponsored by the International Society of Cardiovascular Pharmacotherapy, 36*, 437–447.

Zhivotovsky, B., & Nicotera, P. (2020). *Cell Death and Differentiation, 27*, 2744–2745.

Zhu, C., Hu, W., Wu, H., & Hu, X. (2014). *Sci Rep, 4*, 5029.

CHAPTER TWO

Protein Tyrosine Phosphatase regulation by Reactive Oxygen Species

Colin L. Welsh[a] and Lalima K. Madan[a,b,*]

[a]Department of Cell and Molecular Pharmacology & Experimental Therapeutics, College of Medicine, Medical University of South Carolina, Charleston, SC, United States
[b]Hollings Cancer Center, Medical University of South Carolina, Charleston, SC, United States
*Corresponding author. e-mail address: madanl@musc.edu

Contents

1. Introduction   46
2. The Protein Tyrosine Phosphatase (PTP) catalytic domain   47
3. A Cys-based catalytic mechanism and sensitivity to oxidation   50
4. Sources of Reactive Oxygen Species (ROS) in cellular signaling   53
5. Oxidation of PTPs I: Cellular mechanisms   56
6. Oxidation of PTPs II: Structural aspects   59
7. Conclusions   66
Acknowledgments   67
Disclosure of potential conflicts of interest   67
References   67

Abstract

Protein Tyrosine Phosphatases (PTPs) help to maintain the balance of protein phosphorylation signals that drive cell division, proliferation, and differentiation. These enzymes are also well-suited to redox-dependent signaling and oxidative stress response due to their cysteine-based catalytic mechanism, which requires a deprotonated thiol group at the active site. This review focuses on PTP structural characteristics, active site chemical properties, and vulnerability to change by reactive oxygen species (ROS). PTPs can be oxidized and inactivated by H_2O_2 through three non-exclusive mechanisms. These pathways are dependent on the coordinated actions of other H_2O_2-sensitive proteins, such as peroxidases like Peroxiredoxins (Prx) and Thioredoxins (Trx). PTPs undergo reversible oxidation by converting their active site cysteine from thiol to sulfenic acid. This sulfenic acid can then react with adjacent cysteines to form disulfide bonds or with nearby amides to form sulfenyl-amide linkages. Further oxidation of the sulfenic acid form to the sulfonic or sulfinic acid forms causes irreversible deactivation. Understanding the structural changes involved in both reversible and irreversible PTP

Advances in Cancer Research, Volume 162
ISSN 0065-230X, https://doi.org/10.1016/bs.acr.2024.05.002
Copyright © 2024 Elsevier Inc. All rights are reserved, including those for text and data mining, AI training, and similar technologies.

45

oxidation can help with their chemical manipulation for therapeutic intervention. Nonetheless, more information remains unidentified than is presently known about the precise dynamics of proteins participating in oxidation events, as well as the specific oxidation states that can be targeted for PTPs. This review summarizes current information on PTP-specific oxidation patterns and explains how ROS-mediated signal transmission interacts with phosphorylation-based signaling machinery controlled by growth factor receptors and PTPs.

1. Introduction

Cellular signaling events during proliferation, adhesion, and migration are initiated and transmitted through the phosphorylation of protein tyrosines (Hunter, 1995). Two families of signaling enzymes—Protein tyrosine Kinases (PTKs), which phosphorylate tyrosine residues, and Protein Tyrosine Phosphatases (PTPs), which reverse them—balance this reversible post-translational modification of target proteins (Ahuja, 2018; Almo et al., 2007; Chen, Dixon, & Manning, 2017; Manning, Whyte, Martinez, Hunter, & Sudarsanam, 2002). Approximately 95 PTKs are counteracted by 107 PTPs such that mutations in both are linked to many human diseases including cancers (Andersen et al., 2004; Cohen, 2001). While PTPs were frequently dismissed as housekeeping enzymes that simply counteract the action of PTKs, it is now recognized that PTPs can have both negative and positive regulatory impacts on the signaling cascades (Welsh, Allen, & Madan, 2023; Welsh, Pandey, & Ahuja, 2021). PTPs have recently regained reach as pharmacological targets for a variety of malignancies, thanks to increased patient data availability and clever designing of allosteric inhibitors (Welsh et al., 2023; Zhang, 2017). These efforts are also supported by recent developments in mass spectrometry and peptide chemistry, as well as the convenience of working with purified PTP catalytic domains. PTP regulation occurs at the level of gene expression, through protein–protein interactions, allosteric intradomain interactions, and/or through phosphorylation of their serine/threonine or tyrosine residues (Ahuja, 2018; Ahuja & Gopal, 2014; Welsh et al., 2021). In recent years, oxidation–reduction events, also known as redox reactions, have emerged as a significant mechanism for the regulation of cellular PTP activity by chemical agents (Boivin & Tonks, 2015; Hay et al., 2020; Londhe et al., 2020; Ostman, Frijhoff, Sandin, & Böhmer, 2011; Tonks, 2005, 2006, 2013; Tsutsumi et al., 2017). These reactions are centered on the catalytic cysteine residue of these enzymes. This review briefly covers details on PTP oxidation and its regulation in response to oxidative stress and is intended to update and complement other reviews that have been published on this subject in the past (Boivin & Tonks, 2015; Chiarugi & Cirri, 2003; Netto & Machado, 2022; Tanner, Parsons, Cummings, Zhou, & Gates, 2011).

2. The Protein Tyrosine Phosphatase (PTP) catalytic domain

PTPs are the largest protein phosphatase enzyme superfamily and are mechanistically split into four types (Ahuja, 2018; Chen et al., 2017). Classes I–III contain cysteine-based PTPs with a nucleophilic cysteine in the signature active site $HC-X_5-R$ motif (Zhang, Wang, & Dixon, 1994). The evolutionarily variable Class IV PTPs Eya (Eyes Absent) are related to Haloacid Dehalogenases (HAD) and use aspartate instead of cysteine at the enzyme active site (Burroughs, Allen, Dunaway-Mariano, & Aravind, 2006). Cysteine-based Class I has classical PTP domains and dual-specificity DUSPs. The domain architecture of class I PTPs is highly conserved, and these are strictly selective for pTyr substrates while DUSPs have wider substrate specificity. About 37 classical PTP genes in humans code for 35 active PTP enzymes (Andersen et al., 2001). Furthermore, based on their cell location, traditional PTPs can be split into two groups: membrane-anchored receptor protein tyrosine phosphatases (RPTPs) and intracellular PTPs (Fig. 1A).

PTP1B, the first PTP isolated from the human placenta and examined for its molecular features by protein crystallization, exemplifies the conserved PTP catalytic domain (Barford, Flint, & Tonks, 1994; Tonks, 2003; Tonks, Diltz, & Fischer, 1988). Fig. 1B shows that the catalytic domain's core is made up of twisted β-sheets surrounded by α-helices. Andersen et. al, have defined ten motifs that define the sequence conservation on the defined domain architecture (Fig. 1B) (Andersen et al., 2001). Of these, four motifs include the PTP active site loops while the other six motifs maintain the domain's structural integrity. The Cα-regiovariation score analysis reveals that the most conserved motif is the structural motif-4 (F/Y) IAxQGP, which forms the hydrophobic core around the PTP loop. Motif-4 and motif-3 DYINA(N/S) form a parallel-antiparallel β-sheet at the center of the PTP domain. The PTP domain's hydrophobic cluster also comprises residues from motif-2 DxxR(V/I)xL, motif-5 TxxDFWx(M/L/V)x(W)(E/Q), motif-6 (I/L/V) (V/I)MxT, and motif-7 KCxxYWP. The aromatic side chains of residues from motif 5 and

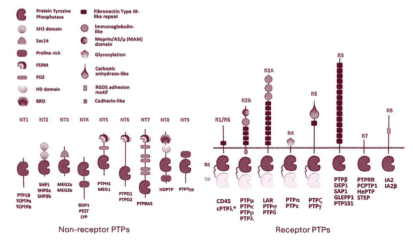

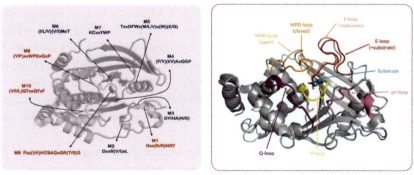

Fig. 1 The PTP Superfamily of Enzymes: (A) Illustration of human PTPs with membrane-bound (receptor) and cytosolic (non-receptor) subtypes. (B) (right) Domain architecture of the conserved PTP domain, indicating the positions of ten sequence motifs that are conserved across the superfamily. (left) An alternative view of the domain showing the active site's position between numerous mobile loops. PTP1B crystal structures have been used for both the illustrations; open structure PDB ID: 5K9V, closed structure PDB ID: 1PTV.

motif 7 (Phe95, Trp96, Tyr124, and Trp125 in PTP1B) help stabilize the core of the PTP domain through π-stacking interactions. Because a protein's hydrophobic core is responsible for its thermostability, mutations in these motifs has a significant impact on the PTP's folding properties (Muise, Vrielink, Ennis, Lemieux, & Tremblay, 1996).

The PTP active site is encased within numerous loops that contain catalytic residues in four conserved sequence motifs (Fig. 1B). The pTyr-loop, also known as the pTyr-recognition loop or substrate binding loop, is responsible for confining PTP's substrate specificity to phosphorylated tyrosines. This loop is a part of motif-1 Nxx(K/R)NRY, which guards the catalytic site and defines the depth of the active site crevice (Madan & Gopal, 2011). The motif has an 84% conserved tyrosine/phenylalanine that works as a causeway to the active site binding pocket, giving it a depth of ~9 Å. As a result, only a pTyr can access the binding pocket's bottom. Furthermore, this tyrosine/phenylalanine aids in hydrophobic packing of the phosphorylated substrate by making π-π stacking interactions with the substrate's phosphotyrosine residue, facilitating substrate binding into the binding site (Madan & Gopal, 2011). An asparagine (or aspartate) two residues downstream from the pTyr-loop creates a bipartite hydrogen bond with the substrate, stabilizing it in the active site (Peti & Page, 2015). The WPD loop is a part of motif-8 ((Y/F)xxWPDxGxP) and includes the general acid/base aspartate required for catalysis. PTPs with variations at the catalytic aspartate (as seen in the second PTP domains of bidomain RPTPs) are either inactive or have extremely poor catalytic activity (Ahuja & Gopal, 2014; Madan, Goutam, & Gopal, 2012; Madan et al., 2011; Welsh et al., 2021). Mutations in the general acid–base aspartate form a clever method of trapping substrates at the active site (Blanchetot, Chagnon, Dubé, Hallé, & Tremblay, 2005). In the apo (substrate-unbound) form, the WPD–loop alternates between open and closed conformations, whereas in the substrate-bound form, the loop primarily samples a closed conformation. The tryptophan residue in the WPD loop acts as a hinge and mediates the loop's mobility. The P-loop, also called the PTP-loop or phosphate-binding loop, is a component of motif-9 VHCSXGXGR(T/S)G. This contains the PTP's signature $HC-X_5-R$ motif's active site cysteine and invariant arginine. The invariant arginine of this motif, combined with the P-loop's backbone nitrogen atoms, generates a strong positively charged milieu in the active site and increases the PTP's affinity for negatively charged phosphotyrosine substrates (Denu & Dixon, 1998). This arginine also aids in the proper placement of the substrate in the active site and stabilizes the cysteinyl–phosphate intermediate during catalysis. The histidine, serine/threonine, and invariant arginine contribute to decreasing the pKa of active site cysteine to around pH 6.5 and allows for it to maintain a deprotonated "thiol" state (Cys-S⁻) (Zhang & Dixon, 1993). Lastly, the Q-loop, which is part of motif-10 ((V/I/L) QTxxQYXF) comprises of two conserved glutamine residues that are essential for PTP active site function. Gln262 and Gln264 restrict the transfer of the phosphoryl group from the phosphoenzyme intermediate to a water molecule rather than other nucleophilic acceptors, preserving PTP's hydrolase activity and preventing them from acting as kinase-like phosphotransferases (Zhang & Dixon, 1993; Zhao, Wu, Noh, Guan, & Zhang, 1998). Also, this loop creates hydrogen bonds with the P-loop and interacts with the aspartate of the WPD-loop to maintain the active site conformation required for maximum catalytic activity.

The E-loop is another conserved motif in PTPs, located close to the PTP-loop and WPD-loop (Fig. 1B). This loop is made up of a 100% conserved glutamate residue and two 85%–90% conserved lysine residues, which play crucial roles in coordinating the dynamics of the WPD loop and organizing the active site (Andersen et al., 2001, 2005). The glutamate residue creates a bipartite hydrogen bond with the arginine of the P-loop, stabilizing the guanidium group in a location that increases its accessibility to the incoming phosphate. One of the conserved lysine residues is located close to the glutamate and interacts with the WPD loop's catalytic aspartate via electrostatic interactions. This helps the WPD-loop to maintain its closed conformation in the substrate-bound state. Indeed, changing this lysine to alanine reduces PTP's catalytic performance significantly. The second lysine residue is located at the rim of the active site and aids in lowering the pKa of catalytic cysteine while also stabilizing the incoming pTyr substrate by giving a surface positive charge. Although the E-loop can take various forms (e.g., β-hairpin, fully disordered), the interaction of glutamate with PTP-loop arginine and the role of the two lysine residues are consistent across all PTPs (Asante-Appiah et al., 2006; Scapin, Patel, Patel, Kennedy, & Asante-Appiah, 2001).

3. A Cys-based catalytic mechanism and sensitivity to oxidation

All PTPs have one catalytic mechanism for phosphate monoester hydrolysis that utilizes their invariant nucleophilic (deprotonated thiol) cysteine and an aspartic acid as the general acid/base. The reaction follows a two-step double-displacement mechanism that makes a covalent intermediate at the PTP active site (Zhang et al. 1994). The pTyr recognition loop (Tyr 46 in PTP1B) and E-loop (Glu 115 in PTP1B) residues help to admit and place the substrate at the active site (Stage I, Fig. 2). A phenylalanine (Phe182 in PTP1B) downstream from the aspartate in the WPD

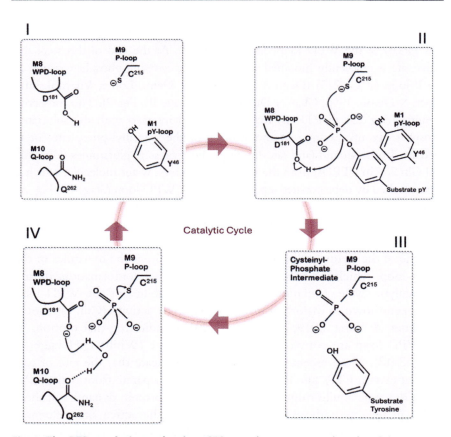

Fig. 2 The PTP catalytic mechanism: PTPs employ a conserved nucleophilic cysteine nested between a conserved HC-X_5-R active site motif. Stage I: The active site cysteine has a low pKa and operates in its thiolate form. Stage II: The pY-loop facilitates the entry of phosphotyrosine into the active site. The cysteine in the active site carries out a nucleophilic attack on the reactive phosphate group. This is facilitated by the aspartate residue in the WPD loop, which acts as a general acid/base. Stage III: A covalent intermediate is formed at the active site in the form of a cysteinyl phosphate group. Stage IV: glutamine from the Q loop activates water molecules to aid hydrolysis of the cysteinyl phosphate intermediate and regenerates the active site for subsequent catalytic cycles.

loop generates π-π stacking contacts with the phosphotyrosine substrate, which helps align the substrate at the active site. In the first chemical step, the invariant cysteine (Cys215 in PTP1B) functions as a nucleophile, attacking the phosphorus on the phosphotyrosine substrate. This triggers the breakage of the phosphorus–oxygen link, resulting in the ejection of the product tyrosine, which is simultaneously protonated by the aspartate (Asp181 in PTP1B) of the WPD-loop (Stage II, Fig. 2). At the end of this step, the active site is covalently modified and contains a cysteinyl-phosphate group at the P-loop (at Cys215) (Denu & Dixon, 1998; Denu, Lohse, Vijayalakshmi, Saper, & Dixon, 1996; Guan & Dixon, 1991) (Stage III, Fig. 2). The conserved arginine (Arg 221 in PTP1B) from the P-loop interacts with the substrate's phosphate group, assisting in the stability of the cysteinyl-phosphate intermediate. In the second (and rate-limiting) chemical step, glutamines (Gln 262, and Gln 266 in PTP1B) from the Q-loop sequester a water molecule, which is then attacked by deprotonated aspartate from the WPD loop (Stage IV, Fig. 2). Simultaneously, catalytic aspartate attacks the intermediate's phosphorus-sulfur link, facilitating hydrolysis of the covalent enzymatic intermediate and the release of free phosphate.

One of the factors influencing PTP catalytic rate is the dynamics of the WPD-loop, which varies between open and closed conformations during the catalytic cycle. Mutations affecting the flexibility of the WPD-loop have been shown to majorly impact PTP catalytic activity (Cui, Lipchock, Brookner, & Loria, 2019; Moise et al., 2018), indicating that motions of the WPD loop are correlated to the reaction rate (Whittier, Hengge, & Loria, 2013). More recent studies, however, indicate that the rate of catalysis of certain PTPs like PTP1B and HePTP is separate from their WPD loop's dynamic equilibrium and that the internal protein dynamics of these PTPs, in regions both around and away from the active site influences catalytic activity (Choy et al., 2017; Crean, Biler, van der Kamp, Hengge, & Kamerlin, 2021; Cui et al., 2019). Therefore, it is more plausible that the process of catalysis is controlled by coherent and coordinated movements inside the active site, either in reaction to the opening of the loop or when the active site is in a completely closed condition (Choy et al., 2017; Torgeson, Clarkson, Kumar, Page, & Peti, 2020).

Oxidized PTPs have been reported in cells at physiological conditions indicating that redox processes underlie signaling events in cells (Tonks, 2005). Evidence suggests that redox signaling is especially prevalent in cancer cells where hyperactive PTKs mediate oxidative inhibition of PTPs (Chiarugi & Cirri, 2003). As the oxidation of cysteine in the active site of PTP results in the formation of an enzyme containing a non-functional catalytic center; it allows for uninterrupted phosphorylation signaling by PTKs in the absence of a counteractive phosphatase action (Chiarugi & Buricchi, 2007). PTP1B is reported to undergo oxidation in MEF thioredoxin $-/-$ cells in response to PTK signaling (Lou et al., 2008). It is then reactivated by the thioredoxin system as is seen in NIH3T3 cells (Markus Dagnell et al., 2013). Oxidation of PTP1B by H_2O_2 is also reported in hepatoma (HepG2), adipose (3T3-L1), and human cancer A431cells in response to insulin receptor signaling (Lou et al., 2008; Mahadev, Zilbering, Zhu, & Goldstein, 2001). PTPs including SHP2, PTEN, DUSP1, and LAR are reported to be oxidized in angiomyolipoma tumor cells as a result of platelet-derived growth factor (PDGF) signaling (Boivin, Zhang, Arbiser, Zhang, & Tonks, 2008). Angiotensin II signaling in vascular smooth muscle cells produces oxidants that hinder SHP2 activity and stimulate Akt signaling (Frijhoff, Dagnell, Godfrey, & Östman, 2013). Oxidization of SHP1 and SHP2 proteins have been reported in EOL-1 myeloid leukemia and human embryonic kidney HEK293 cells (Weibrecht et al., 2007). SHP2 undergoes oxidation in fibroblasts during cell adhesion in response to PDGF stimulation, as well as in T-cells during cell migration in response to T cell receptor signaling (Frijhoff et al., 2013). RPTPε is reported to dimerize and undergo oxidation during EGF signaling and because of oxidative stress induced by treatment of HEK-293 cells with H_2O_2 (Toledano-Katchalski et al., 2003).

4. Sources of Reactive Oxygen Species (ROS) in cellular signaling

All highly reactive oxygen (O_2) derivatives are referred to as Reactive Oxygen Species (ROS) (Bardaweel et al., 2018). This group consists of non-radical peroxides like hydrogen peroxide (H_2O_2), hypochlorous acid (HOCl), and singlet oxygen ($^1O_2$) that all contain a volatile O—O linkage and are therefore prone to becoming sources of free radicals. ROS also includes species of radicals like superoxide ($O_2^{\bullet-}$), hydroxyl ($^{\bullet}OH$), peroxyl ($RO_2^{\bullet}$), and alkoxyl ($RO^{\bullet}$) that have unpaired valence shell electrons (Bedard & Krause, 2007). According to several studies, hydrogen peroxide (H_2O_2) is the dominant ROS agent in cells (Dagnell, Cheng, & Arnér, 2021; Dagnell et al., 2019; Di Marzo, Chisci, & Giovannoni, 2018). The tendency of the highly reactive superoxide ($O_2^{\bullet-}$) to rapidly undergo dismutation, both spontaneously and enzymatically, results in the formation of H_2O_2 which easily diffuses through membranes and across the cytosol. The O—O bond in H_2O_2, although less strong than the bond in dioxygen (O_2), makes the molecule relatively stable compared to other radical species. This stability allows H_2O_2 enough time to react with specific targets that it can oxidize. While H_2O_2 has been considered a toxic molecule since its discovery, in more recent times, it has become evident that the production of low concentrations of H_2O_2, under strict physiological control, plays a significant role in the regulation of several important signaling pathways including those involved in cell proliferation, differentiation, metabolism, and cell migration (Heo, Kim, & Kang, 2020; Lennicke, Rahn, Lichtenfels, Wessjohann, & Seliger, 2015; Prasad, Gupta, & Tyagi, 2017; Sies, 2017; Vilchis-Landeros, Matuz-Mares, & Vázquez-Meza, 2020).

The NADPH oxidase (NOX) family, which was initially identified as a source of microbiocidal oxidants produced in response to lipopolysaccharides, is recognized as the major producer of cellular ROS (Bedard & Krause, 2007; Brown & Griendling, 2009) (Fig. 3A). The family includes seven different kinds of NOX complexes, known as NOX1, NOX2 (gp91phox), NOX3, NOX4, NOX5, DUOX1, and DUOX2 that are widely distributed in various cellular locations including the cell membranes, mitochondria, peroxisomes, and the endoplasmic reticulum. Their catalytic subunit is a transmembrane protein that binds electron-transferring FAD and heme prosthetic groups which can take an electron from NADPH. This electron is then used to reduce molecular O_2, leading to the production of $O_2^{\bullet-}$, that is eventually converted to H_2O_2 by dismutases (Fig. 3A). Research suggests that NOX -dependent ROS generation is linked with oncogenic signaling of RAS (Choi, Kim, Song, Kim, & Kim, 2008) and various growth factors including TGF-b1, interleukin-1, TNFa, insulin, PDGF, EGF, angiotensin II, thrombin, and lysophosphatidic acid (Mahadev et al., 2001; Mesquita et al., 2010; Ohba, Shibanuma, Kuroki, & Nose, 1994; Sundaresan, Yu, Ferrans, Irani, & Finkel, 1995; Svegliati et al., 2005; Weng, Chang, Hung, Yang, & Chien, 2018). Also, growth-factor stimulation activates the PI3K pathway, leading to Rac1 activation and triggers NADPH-oxidase activity (Bae et al., 2000; Svegliati et al., 2005).

ROS are also produced as a byproduct of lipid metabolism. Several oxygenases, including lipoxygenase (Lox) and cyclooxygenase (Cox) enzyme families produce ROS during fatty acid metabolism as well as hormone and inflammatory mediator biosynthesis (Fig. 3B). Eicosanoid biosynthesis occurs by the oxidation of fatty acids, such as arachidonic acid, catalyzed by the LOX enzymes. Alternatively, the COX enzymes can also catalyze the oxidation of fatty acids, resulting in the generation of prostaglandins. Both processes necessitate the introduction of molecular oxygen (O_2) to arachidonic acid, resulting in the generation of ROS molecules as a byproduct of this oxidative process. Additionally, leukotrienes, which are

Protein Tyrosine Phosphatase regulation by Reactive Oxygen Species

A. Intracellular production of ROS

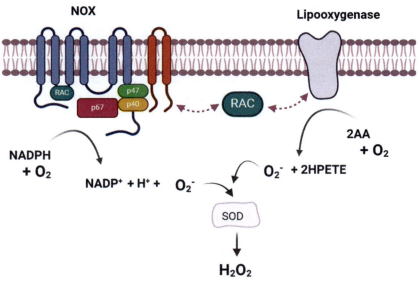

B. Mitochondrial source of ROS

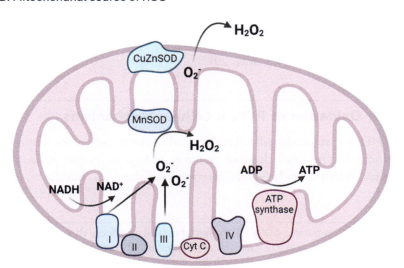

Fig. 3 Intracellular sources of ROS: (A) NADPH oxidase (NOX) enzymes are the primary producers of ROS. NOX enzymes are localized to several organelles and cellular membranes where they create superoxide ($O_2^{·-}$) radicals as a byproduct of NADPH oxidation. Lipooxygenases that convert arachidonic acid (2AA) to hydroperoxyeicosatetraenoic acids (HPETEs) also release superoxide ($O_2^{·-}$) into the cytosol. Subsequently, these superoxide molecules are transformed into hydrogen peroxide (H_2O_2) by the various superoxide

(Continued)

the final products of the LOX pathways, or intermediates such as hydroperoxyeicosatetraenoic acids (HPETEs) can induce the generation of ROS molecules from NOX and dismutatses (Boivin & Tonks, 2015; Choi et al., 2008).

An unintentional, yet critical source of H_2O_2 is the mitochondrial respiratory chain (Brand, 2016; Mailloux, 2015). ROS are produced as a byproduct of the electron transport chain's activity during cellular aerobic respiration in the mitochondrion when O_2 is prematurely reduced by one electron and forms superoxide ($O_2^{\bullet-}$) (Fig. 3B). The amount of O_2^- that is created during the transfer of electrons from NADH to O_2 and the simultaneous generation of ATP is determined by the relationship between the pool of accessible NADH electron donors, the local O_2 concentration alongside the efficiency of the ATP synthase. $O_2^{\bullet-}$ is converted into membrane-permeable H_2O_2 by the mitochondrial manganese superoxide dismutase (MnSOD) located in the matrix and copper/zinc superoxide dismutase (CuZnSOD) localized in the mitochondrial intermembrane space (Bedard & Krause, 2007). Hydrogen peroxide diffuses through mitochondrial membranes and into the cytosol via specific aquaporins (AQP) called peroxiporins (Sies, 2017). Oxidative stress caused by conditions such as hypoxia, injury to the mitochondria, or decreased levels of these free radical scavenging enzymes, enhance the "leakage" of ROS from the mitochondria and can trigger cell death (Prasad et al., 2017; Reczek & Chandel, 2015).

5. Oxidation of PTPs I: Cellular mechanisms

PTP's cysteine-based active site makes them extremely sensitive to ROS action. The thiol chain of cysteine undergoes oxidation forming a reversible sulfenic acid (SOH) intermediate that can transform into a

---

Fig. 3—Cont'd dismutases (SOD). (B) The mitochondrial respiratory chain is an incidental source of ROS. Superoxide ($O_2^{\bullet-}$) radicals are created as a byproduct of ATP production and the movement of electrons through the electron transport chain. Included in the inner mitochondrial membrane is the respiratory chain complex comprised of NADH dehydrogenase (complex I), the succinate-coenzyme Q reductase complex (complex II), the cytochrome b-c1 complex (complex III), the cytochrome oxidase complex (complex IV) and the ATP synthase complex. Mitochondrial dismutases including MnSOD (manganese superoxide dismutase) and CuZnSOD (copper/zinc superoxide dismutase) convert superoxide ($O_2^{\bullet-}$) radicals into the membrane permeable hydrogen peroxide (H_2O_2).

disulfide bond (S—S) with nearby cysteines or a sulfenyl amide (S—N) bond with a neighboring backbone nitrogen. These can be reversed to the "active" thiol form by antioxidants such as glutathione, thioredoxin, and their associated thiol-transferases like glutaredoxin (Ostman et al., 2011). Further oxidation of the sulfenic acid (SOH) form creates the hyperoxidized sulfinic (SO_2H) and sulfonic (SO_3H) states that are resistant to regeneration and lead to irreversible PTP inactivation (Fig. 4). An increasing number of studies show that ROS are not a simplistic tool for the broad oxidation of random targets. The properties of the oxidizing agents produced have implications for the particular signaling response that is activated. *Juarez et al.* have demonstrated that PTP1B is suppressed by H_2O_2, but not by $O_2^{\bullet-}$ (Juarez et al., 2008). Similarly, Choi et al. show that specificity at the level of oxidant-targeting exhibits preferences for the target's downstream effectors (Choi, Kim, Jhon, & Lee, 2011). Their studies how that SHP1 is inactivated in hematopoietic progenitors by ROS production caused by macrophage colony stimulating factor, but its close homologue SHP2 is not affected. Also, the oxidative inactivation of SHP1

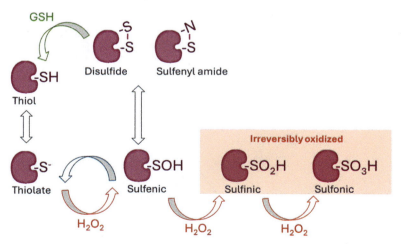

Fig. 4 PTP oxidation pathways: In the presence of H_2O_2, the PTP catalytic cysteine's thiolate ion is reversibly oxidized to sulfenic acid. Sulfenic acid's instability causes it to quickly transform into a thiolate ion or reversibly oxidized intramolecular disulfide or sulfenyl amide state. Agents like glutathione and thioredoxin can reduce these to a thiol group. However, further oxidation of the sulfenic acid by H_2O_2 creates irreversibly oxidized sulfinic acid and sulfonic acid forms that permanently inactivate PTP active sites.

specifically impacts the PI3K/Akt survival pathway, while the mitogen-activated protein kinase (MAPK) survival signaling remains unaffected, despite both pathways being regulated by SHP1.

Kinetics wise, oxidation of PTPs with H_2O_2 is much slower ($\sim 10^1$ M^{-1} s^{-1}) compared to other reactive cysteine containing enzymes like Peroxiredoxins (Prxs) ($\sim 10^7$ M^{-1} s^{-1}) (Winterbourn & Metodiewa, 1999). These Prx enzymes serve as sensors of ROS in cells and react specifically with hydroperoxides, but not with other electrophiles such as iodoacetamides or chloroamines (Peskin et al., 2007). Moreover, as peroxiredoxins are highly abundant in cells, and have a faster rate of oxidation, it is understood that they outcompete PTPs for reaction with H_2O_2. At present three non-mutually exclusive mechanisms of PTP oxidation have been outlined (Fig. 5) (Netto & Antunes, 2016). The first model is of direct reaction of H_2O_2 with the PTP active site cysteine. The second model details an indirect, yet specific, oxidation method following a "redox relay mechanism" involving sequential transfers of oxidizing equivalents between Prxs and PTPs. At first, catalytic cysteines on Prx enzymes are oxidized by H_2O_2 to sulfenic acids (CysSOH) that subsequently are converted to intra- or intermolecular disulfides (Cys-SS-Cys). Oxidized Prx enzymes use thiol-disulfide exchange reactions to transfer oxidizing equivalents to PTPs that they

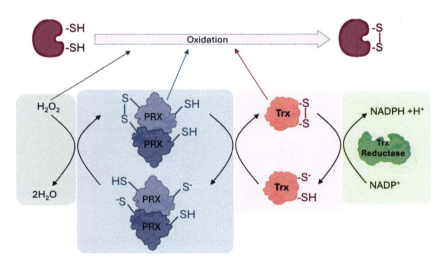

Fig. 5 Cellular mechanisms of PTP oxidation: PTPs are oxidized by H_2O_2 directly, or indirectly through the oxidation of peroxiredoxin (Prx) and thioredoxin (Trx) proteins that contain reactive thiols with higher affinity for hydrogen peroxide (H_2O_2) than PTPs. In the indirect mechanism, a redox relay of thiol-disulfide exchange reactions engages several proteins that are eventually reduced by glutathione and reductases.

select by specific protein-protein interactions (Marinho, Real, Cyrne, Soares, & Antunes, 2014; Netto & Antunes, 2016; Perkins, Nelson, Parsonage, Poole, & Karplus, 2015). Support for this model comes from studies on Ras-induced breast cancer in mammary epithelial cells where Prx1 is seen to be critical for supporting the tumor suppressive function of PTEN phosphatase (Cao et al., 2009). Here, physical interaction between Prx1 and PTEN, and thiol exchange between the two proteins is reported to be critical for protecting PTEN from oxidation-induced inactivation. The third model of H_2O_2 sensing includes thioredoxin (Trx) enzymes that also function as redox sensors and can perform the "redox relay" with either PTPs or Prx enzymes. Studies show that while Trx is a slower redox sensor than Prx, it is extremely selective in its protein-protein interactions under different redox conditions (Berndt, Lillig, & Holmgren, 2007; Palde & Carroll, 2015). Subtle structural changes in Trx modify its protein interaction surfaces such that certain Trx-interactor complexes only exist when Trx is in its oxidized form (Berndt et al., 2007; Psenakova et al., 2020). For example, in vitro- and cell-based assays indicate that Trx1 binds and reduces oxidized PTP1B protein but not the SHP2 phosphatase (Dagnell et al., 2013). Cells lacking Trx reductase TrxR1 (that regenerates reduced Trx1) show increased oxidation of PTP1B with no corresponding change in SHP2 oxidation. Hence models 2 and 3 indicate that redox sensing by PTPs is specified by critical protein-protein interactions that are a maneuver around the slow direct oxidation rates of these proteins. Under high H_2O_2 concentrations when Prxs are hyperoxidized to sulfinic or sulfonic states, localized accumulation of H_2O_2 allows for direct oxidation of PTPs despite their low oxidation rates. This is called the "floodgate hypothesis" (Rhee & Woo, 2011). However, this hypothesis is controversial because several studies show Prxs to be much more abundant than PTPs (Perkins et al., 2015), and that they use several alternate pathways for rapid regeneration (Rhee & Woo, 2011). Also, when Prxs are permanently inactivated by oxidation, cells have other antioxidants like GSH and GAPDH that can easily outcompete PTPs for chemical reactions with H_2O_2 (Winterbourn & Hampton, 2008).

6. Oxidation of PTPs II: Structural aspects

Crystal structures of several PTPs have allowed for a thorough analysis of their H_2O_2 concentration-dependent oxidation (Table 1). PTP1B has been crystallized in all oxidized states of its catalytic cysteine

Table 1 List of crystal structures of PTPs exhibiting their active sites cysteine in different oxidized states.

Protein name	Active site cysteine	PDB ID	References
PTP1B	Reduced	5K9V	Choy et al. (2017)
PTP1B	Oxidized to sulfenyl amide	1OEM	Salmeen et al. (2003)
PTP1B	Oxidized to sulfenyl amide	1OES	Van Montfort et al. (2003)
PTP1B	Oxidized to sulfonic acid	1OEO	Salmeen et al. (2003)
PTP1B	Oxidized to sulfonic acid	1OET	Van Montfort et al. (2003)
PTP1B	Oxidized to sulfenic acid	1OEU	Van Montfort et al. (2003)
PTP1B	Oxidized to sulfinic acid	1OEV	Van Montfort et al. (2003)
PTPRU D1 domain	Reduced	6SUB	Hay et al. (2020)
PTPRU D1 domain	Oxidized to sulfonic acid	6SUC	Hay et al. (2020)
PTPσ D1D2	D1 oxidized to sulfenic acid, D2 reduced	4BPC	Jeon et al. (2013)
LYP	Reduced	2QCJ	Unpublished, structure available in the PDB
LYP	Oxidized, disulfide bonded to backdoor cysteine	3H2X	Tsai et al. (2009)
SHP2 (N308D mutant)	Reduced	4NWF	Unpublished, structure available in the PDB
SHP2 (N308D mutant)	Oxidized, disulfide bonded to backdoor cysteine	6ATD	Machado, Critton, et al. (2017)
STEP	Modified to S-ACETYL-CYSTEINE	2BV5	Eswaran et al. (2006)

(Fig. 6) (Salmeen et al., 2003; van Montfort, Congreve, Tisi, Carr, & Jhoti, 2003). The reversible sulfenyl amide form consists of a covalent bond between the active site cysteine and the backbone nitrogen of the neighboring serine residue in the P-loop. Initially, it was suggested that sulfenyl amide formation in PTP1B proceeded via direct S_N2 mechanism when the backbone nitrogen of the Ser216 nucleophilically attacked the Sγ atom of the active site Cys215 SOH present in the sulfenic acid form (van Montfort et al., 2003). This was supported by the protonation of the active site His214 which also maintains the thiol form of Cys215 (Zhang & Dixon, 1993). More recently, molecular dynamics simulations and high-level hybrid quantum mechanics/molecular mechanics-based studies show that sulfenyl amide formation occurs through an iminol-type intermediate that requires the participation of Glu115 (of the E loop) alongside His214 (Dokainish & Gauld, 2015).

In the sulfenyl amide state, the active site of PTP1B is distorted and shows alterations in the positioning of its active site loops (Salmeen et al., 2003; van Montfort et al., 2003). Changes are most pronounced in the conformational of the pY loop such that Tyr46 residue flips and breaks its interactions with the Ser 216 of the P-loop. Gly218 of the P-loop pivots by around 7 Å, and the P-loop switches to break interactions between the side chain of Ser216 and Glu115 of the E-loop. Gln262 of the Q-loop switches its conformations and faces away from the active site (Fig. 6). These perturbations expose Ser50 to the cytosol and promote its phosphorylation by AKT (Ravichandran, Chen, Li, & Quon, 2001). The oxidized and phosphorylated PTP1B is transiently stabilized by binding of 14–3-3ζ protein and promotes sustained EGFR signaling as is seen in HEK293 cells (Londhe et al., 2020). Distortion in the active site of PTP1B in the sulfenyl amide form has been used to generate conformation-specific antibodies that can allow for studying reversible oxidation of PTP1B in cellular systems (Haque, Andersen, Salmeen, Barford, & Tonks, 2011). Krishnan et al. have used intrabody scFv45 to trap the reversibly oxidized PTP1B-OX in the catalytically inactive form and studied its effect on enhancing insulin and leptin signaling in hepatic stellate cells (Krishnan et al., 2018). These studies reveal the intricate interplay between phosphorylation and redox-based signaling. Efforts are underway to harness the intrabody approach as a novel therapeutic methodology for the treatment of diabetes and obesity.

When compared to the sulfenyl amide form, perturbations in PTP1B active site aren't as pronounced in the higher oxidized sulfenic, sulfinic, and sulfonic acid states (Salmeen et al., 2003; van Montfort et al., 2003) (Fig. 6).

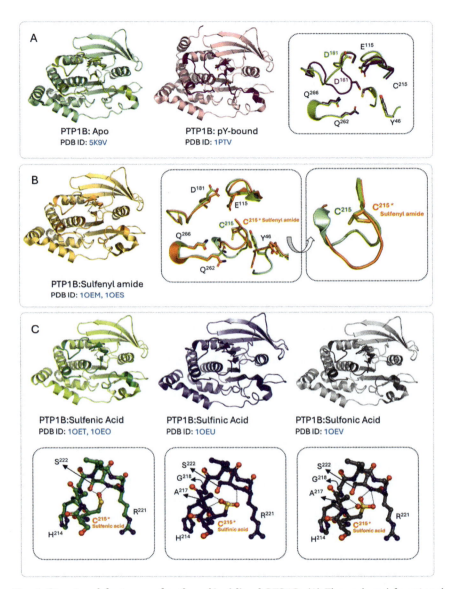

Fig. 6 Structural features of reduced/oxidized PTP1B: (A) The reduced functional PTP1B contains a central active site Cysteine in a Thiol form flanked by various other active site loops. The WPD loop faces away from the active site cysteine in the apo (left, *green*) and moves towards the cysteine when the substrate binds (center, *pink*). The inset on the right shows the relative movements of the active site loops upon phosphotyrosine binding. (B) The reversibly oxidized sulfenyl amide form of PTP1B is formed when the active site cysteine makes an "S-N" bond with the backbone

In these states, the active site loops maintain the overall conformation as seen in the reduced Cys215 of the active enzyme. In the sulfenic form, Cys215 SOH forms a hydrogen bond with the side chain and backbone amide of Ser222 and keeps the P-loop locked in a tight conformation. Similarly, in the sulfinic form, Cys215 SO_2H forms hydrogen bonds with Ser222 that are stabilized by backbone amide groups of Ala217 and Gly218, also of the P-loop. Lastly, in the sulfonic form, Cys215 SO_3H forms similar hydrogen-bonded interactions as seen in the sulfenic and sulfinic acid states. Sulfenic and sulfonic acid states have also been crystallized for the receptor PTPσ where its active site Cys1589 shows similar conformations as it were in the reduced active form (Table 1) (Jeon, Chien, Chun, & Ryu, 2013). Like in PTP1B, higher oxidized forms of PTPσ show an open conformation of its WPD-loop. Interestingly, PTPσ has two cytosolic PTP domains, but only its membrane proximal D1-PTP domain is oxidized while its membrane distal D2-PTP domain remains reduced. As the D2-PTP domain is a naturally occurring pseudo-enzyme (lacks phosphatase activity), its differential oxidation/reduction processes from the D1 domain suggest an allosteric interplay between the two PTP domains of PTPσ.

Some PTPs including LYP, SHP1/SHP2, STEP, PTPRR, and PTPρ use a conserved cysteine in their motif 7 (K**C**xxYWP) to make a reversible disulfide bond with their active site cysteines (Chen, Willard, & Rudolph, 2009; Cunnick, Dorsey, Mei, & Wu, 1998; Hay et al., 2020; Machado, Critton, Page, & Peti, 2017; Machado, Shen, Page, & Peti, 2017; Tsai et al., 2009). This partnering cysteine here is referred to as the "backdoor" cysteine because of its location in the backside of the PTP active site. Disulfide formation does not alter the conformation of the P-loop but is mediated by a simple rotation of the active cysteine to access the backdoor cysteine (Fig. 7). In LYP, a disulfide bond between the active site Cys227 and backdoor Cys129 stabilizes a partially open active site with the WPD-loop in an intermediate conformation (Tsai et al., 2009). This intermediate conformation is stabilized by electrostatic interactions between Glu133 of

---

nitrogen atom of the neighboring serine. This structure shows a unique and distinct conformational change in its P-loop (inset, right). (C) Higher oxidized forms of PTP1B include its cysteine's oxidation to sulfenic acid (green, left), sulfinic acid (center, blue), and sulfonic acid (right, gray). The sulfinic and sulfonic acid forms cannot be reduced by antioxidants to regenerate the active form of the enzyme. The insets under each structure show the chemical groups of the P-loop and highlight the hydrogen bonds between the active site cysteine and its neighboring residues.

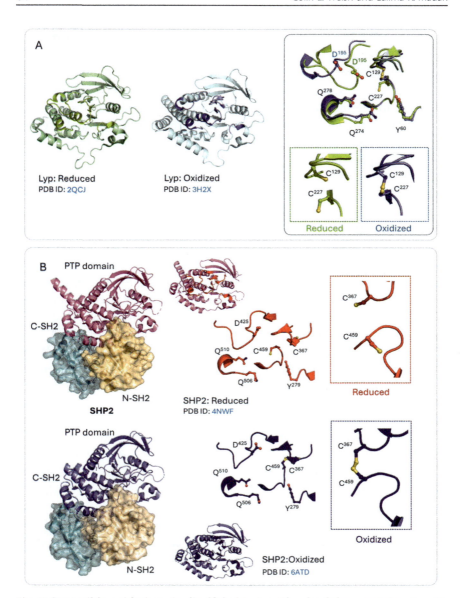

Fig. 7 Reversible oxidation via disulfide binds with a backdoor cysteine: Lyp (A, top) and SHP2 (B, bottom) use the backdoor cysteine in motif 7 (KCxxYWP) to make a reversible disulfide bond and prevent irreversible oxidation of the active site cysteine to sulfinic and sulfonic acids. The insets on the right show the active site structure and disulfide bond formation in the two PTPs.

the E-loop and Arg233 of the P-loop. Interestingly, LYP is proposed to have two more backdoor cysteines including Cys139 in the E-loop and Cys231 that is downstream of the active site cysteine in the P-loop. It is unclear how the active site Cys227 chooses to form disulfide bonds given that the three backdoor cysteines are easily accessible. Biochemical data shows that Cys129S and Cys231S mutants of LYP allow for reversible oxidation of the PTP, perhaps by engaging the third Cys139 from the E-loop(Tsai et al., 2009).

In SHP2, the active site Cys459 engages the backdoor Cys367 in a disulfide bond, and its WPD-loop is seen in an intermediate conformation (Machado, Critton, et al., 2017) (Fig. 7). In both SHP1 and SHP2, a conserved proline in motif 4 (YILTQG**P**) is naturally seen as a cysteine (Cys327 in SHP1 and Cys333 in SHP2) that functions as the second backdoor cysteine and supports reversible oxidation of SHP1/SHP2 (Chen et al., 2009). Removal of both backdoor cysteines (Cys327, Cys373 in SHP1 and Cys333, Cys367 in SHP2) is necessary for the irreversible oxidative inactivation of SHP1/SHP2 to sulfinic and sulfonic acid states. Mutations of the second backdoor cysteine to the canonically conserved proline residues enhance the thermal stability of SHP1/SHP2 catalytic domains (Yarnall, Kim, Korntner, & Bishop, 2022). As SHP1/SHP2 is allosterically regulated by the movements in their tandem SH2 domains (Welsh et al., 2023), the presence of a second backdoor cysteine hints towards an evolved redox protection of the PTPs at the cost of protein stability.

Given that the backdoor cysteine is a conserved residue in motif 7 (KCxxYWP) of all PTPs, it's intriguing that only some of them employ it for an intramolecular disulfide bond. Furthermore, the extent of antioxidant protection afforded by the backdoor cysteine differs among related PTPs. Mutation of the backdoor cysteines to serine in STEP (Cys384) and PTPRR (Cys501) makes them more sensitive to H_2O_2 mediated oxidation when compared to the wild-type proteins, but PTPRR-C501S is less sensitive than STEP-C384S mutant and more easily reactivated by chemical reducing agents such as Dithiothreitol (DTT) (Machado, Shen, et al., 2017). Perplexingly, these studies also report that in the closely related HePTP, the backdoor cysteine Cys183 is dispensable for its reversible oxidation. Unlike its related STEP and PTPRR proteins (collectively known as the Kinase Interaction Motif (KIM)-PTPs), which dephosphorylate mitogen–activated protein kinases (MAPKs), HePTP is reported to form intermolecular disulfide bonds between two protein monomers; using the active site Cys270 of one protomer and a surface exposed Cys116 on another protomer (Machado, Shen, et al., 2017). Also, the reactivation rates of HePTP by DTT were found to be like those observed for PTP1B and SHP1/SHP2, although STEP and PTPRR were reported to be extremely slowly reactivated by reducing agents. These findings imply that protein-specific, most likely dynamics-based indicators may play a role in determining each PTP's oxidative proclivity and reactivation susceptibility.

Despite all the sequence and structural data available, it is impossible to predict how a PTP would react to or protect itself against oxidation. Nonetheless, the conserved backdoor cysteine has been explored as a plausible allosteric site that could be harnessed for PTP inhibition. Covalent modification of PTP1B's backdoor Cys121 by 4-(aminosulfonyl)-7-fluoro-2,1,3-benzoxadiazole (ABDF) is reported to decrease its catalytic activity by several fold without affecting peptide substrate binding (Hansen et al., 2005). ABDF has also been shown to interact with and inhibit the catalytic activity of TC-PTP and LAR PTPs while having no effect on CD45, indicating some selectivity. ABDF treatment of CHO-hIR cells has been shown to suppress PTP1B activity and allow for sustained insulin receptor phosphorylation and signaling. Similarly, covalent modification of Cys121 by 1,2-naphthoquinone is reported to decrease PTP1B's catalytic activity and allow for persistent transactivation of EGFR in human epithelial A431 cells (Iwamoto et al., 2007). More recently, biconjugate PTP1B inhibitors have been designed to covalently modify Cys121 while simultaneously engaging the active site Cys215 in a non-covalent interaction (Khan, Bjij, & Soliman, 2019; Punthasee et al., 2017). These use electrophilic α-bromoacetamide or α-fluoroacetamide groups conjugated to 5-aryl-1,2,5-thiadiazolidin-3-one 1,1-dioxide inhibitors that selectively react with the distal Cys121 and provide for an allosteric mechanism of PTP1B inhibition.

7. Conclusions

Oxidation of PTPs adds another complexity to the already intricate landscape of phosphorylation and redox-based cellular signaling. Currently, there is a renewed interest in understanding the structural and pharmacological processes that regulate PTPs to target them for drug discovery. Understanding the protein-specific reversibly oxidized states of PTPs is crucial for accessing and targeting these proteins. Conformation-specific characteristics of oxidized PTP1B have been utilized to produce antibodies that may help target it for therapeutic intervention, but several key mechanistic concerns remain unexplained. For example, why does the active site cysteine in PTP1B form a sulfenyl amide bond with the adjacent serine rather than a disulfide link with its backdoor cysteine? What chemical characteristics of the active site and backdoor cysteines influence their interactions in a reversible oxidized state? The range of oxidized states observed in PTPs suggests a protein-specific preference for oxidation states, which may be connected to protein dynamics rather than conserved sequence and structure. Further research, particularly on the significance of protein dynamics in conjunction with PTP oxidation, is essential to answer these questions.

Acknowledgments

This work is supported by the SC COBRE in Antioxidants and Redox Signaling of the National Institute of General Medical Sciences (NIGMS) (Grant number: 1P30GM140964) and SCTR NIH/NCATS (Grant Number: UL1TR001450) to LKM.

Disclosure of potential conflicts of interest

No potential conflicts of interest are disclosed by the authors.

References

Ahuja, L. G. (2018). *Protein tyrosine phosphatases: Structure, signaling and drug discovery/Lalima G. Ahuja.* Berlin: De Gruyter.

Ahuja, L. G., & Gopal, B. (2014). Bi-domain protein tyrosine phosphatases reveal an evolutionary adaptation to optimize signal transduction. *Antioxidants & Redox Signaling, 20*(14), 2141–2159. https://doi.org/10.1089/ars.2013.5721.

Almo, S. C., Bonanno, J. B., Sauder, J. M., Emtage, S., Dilorenzo, T. P., Malashkevich, V., ... Burley, S. K. (2007). Structural genomics of protein phosphatases. *Journal of Structural and Functional Genomics, 8*(2-3), 121–140. https://doi.org/10.1007/s10969-007-9036-1.

Andersen, J. N., Del Vecchio, R. L., Kannan, N., Gergel, J., Neuwald, A. F., & Tonks, N. K. (2005). Computational analysis of protein tyrosine phosphatases: Practical guide to bioinformatics and data resources. *Methods, 35*(1), 90–114. https://doi.org/10.1016/j.ymeth.2004.07.012.

Andersen, J. N., Jansen, P. G., Echwald, S. R. M., Mortensen, O. H., Fukada, T., Del Vecchio, R., ... Møller, N. P. H. (2004). A genomic perspective on protein tyrosine phosphatases: Gene structure, pseudogenes, and genetic disease linkage. *The FASEB Journal, 18*, 8–30. https://doi.org/10.1096/fj.02-1212rev.

Andersen, J. N., Mortensen, O. H., Peters, G. H., Drake, P. G., Iversen, L. F., Olsen, O. H., ... Møller, N. P. H. (2001). Structural and evolutionary relationships among protein tyrosine phosphatase domains. *Molecular and Cellular Biology, 21*, 7117–7136. https://doi.org/10.1128/mcb.21.21.7117-7136.2001.

Asante-Appiah, E., Patel, S., Desponts, C., Taylor, J. M., Lau, C., Dufresne, C., ... Scapin, G. (2006). Conformation-assisted inhibition of protein-tyrosine phosphatase-1B elicits inhibitor selectivity over T-cell protein-tyrosine phosphatase. *Journal of Biological Chemistry, 281*, 8010–8015. https://doi.org/10.1074/jbc.M511827200.

Bae, Y. S., Sung, J. Y., Kim, O. S., Kim, Y. J., Hur, K. C., Kazlauskas, A., & Rhee, S. G. (2000). Platelet-derived growth factor-induced H(2)O(2) production requires the activation of phosphatidylinositol 3-kinase. *Journal of Biological Chemistry, 275*(14), 10527–10531. https://doi.org/10.1074/jbc.275.14.10527.

Bardaweel, S. K., Gul, M., Alzweiri, M., Ishaqat, A., HA, A. L., & Bashatwah, R. M. (2018). Reactive oxygen species: The dual role in physiological and pathological conditions of the human body. *The Eurasian Journal of Medicine, 50*(3), 193–201. https://doi.org/10.5152/eurasianjmed.2018.17397.

Barford, D., Flint, A., & Tonks, N. (1994). Crystal structure of human protein tyrosine phosphatase 1B. *Science, 263*, 1397–1404. https://doi.org/10.1126/science.8128219.

Bedard, K., & Krause, K. H. (2007). The NOX family of ROS-generating NADPH oxidases: Physiology and pathophysiology. *Physiological Reviews, 87*(1), 245–313. https://doi.org/10.1152/physrev.00044.2005.

Berndt, C., Lillig, C. H., & Holmgren, A. (2007). Thiol-based mechanisms of the thioredoxin and glutaredoxin systems: Implications for diseases in the cardiovascular system. *The American Journal of Physiology-Heart and Circulatory Physiology, 292*(3), H1227–H1236. https://doi.org/10.1152/ajpheart.01162.2006.

Blanchetot, C., Chagnon, M., Dubé, N., Hallé, M., & Tremblay, M. L. (2005). Substrate-trapping techniques in the identification of cellular PTP targets. *Methods, 35*(1), 44–53. https://doi.org/10.1016/j.ymeth.2004.07.007.

Boivin, B., & Tonks, N. K. (2015). PTP1B: Mediating ROS signaling to silence genes. *Molecular & Cellular Oncology, 2*(2), e975633. https://doi.org/10.4161/23723556.2014.975633.

Boivin, B., Zhang, S., Arbiser, J. L., Zhang, Z.-Y., & Tonks, N. K. (2008). A modified cysteinyl-labeling assay reveals reversible oxidation of protein tyrosine phosphatases in angiomyolipoma cells. *Proceedings of the National Academy of Sciences, 105*(29), 9959–9964. https://doi.org/10.1073/pnas.0804336105.

Brand, M. D. (2016). Mitochondrial generation of superoxide and hydrogen peroxide as the source of mitochondrial redox signaling. *Free Radical Biology and Medicine, 100*, 14–31. https://doi.org/10.1016/j.freeradbiomed.2016.04.001.

Brown, D. I., & Griendling, K. K. (2009). Nox proteins in signal transduction. *Free Radical Biology and Medicine, 47*(9), 1239–1253. https://doi.org/10.1016/j.freeradbiomed.2009.07.023.

Burroughs, A. M., Allen, K. N., Dunaway-Mariano, D., & Aravind, L. (2006). Evolutionary genomics of the HAD superfamily: Understanding the structural adaptations and catalytic diversity in a superfamily of phosphoesterases and allied enzymes. *Journal of Molecular Biology, 361*, 1003–1034. https://doi.org/10.1016/j.jmb.2006.06.049.

Cao, J., Schulte, J., Knight, A., Leslie, N. R., Zagozdzon, A., Bronson, R., ... Neumann, C. A. (2009). Prdx1 inhibits tumorigenesis via regulating PTEN/AKT activity. *The EMBO Journal, 28*(10), 1505–1517. https://doi.org/10.1038/emboj.2009.101.

Chen, C.-Y., Willard, D., & Rudolph, J. (2009). Redox regulation of SH2-domain-containing protein tyrosine phosphatases by two backdoor cysteines. *Biochemistry, 48*(6), 1399–1409. https://doi.org/10.1021/bi801973z.

Chen, M. J., Dixon, J. E., & Manning, G. (2017). Genomics and evolution of protein phosphatases. *Science Signaling, 10*(474), eaag1796. https://doi.org/10.1126/scisignal.aag1796.

Chiarugi, P., & Buricchi, F. (2007). Protein tyrosine phosphorylation and reversible oxidation: Two cross-talking posttranslation modifications. *Antioxidants & Redox Signaling, 9*(1), 1–24. https://doi.org/10.1089/ars.2007.9.1.

Chiarugi, P., & Cirri, P. (2003). Redox regulation of protein tyrosine phosphatases during receptor tyrosine kinase signal transduction. *Trends in Biochemical Sciences, 28*(9), 509–514. https://doi.org/10.1016/S0968-0004(03)00174-9.

Choi, H. K., Kim, T. H., Jhon, G. J., & Lee, S. Y. (2011). Reactive oxygen species regulate M-CSF-induced monocyte/macrophage proliferation through SHP1 oxidation. *Cell Signaling, 23*(10), 1633–1639. https://doi.org/10.1016/j.cellsig.2011.05.017.

Choi, J. A., Kim, E. Y., Song, H., Kim, C., & Kim, J. H. (2008). Reactive oxygen species are generated through a BLT2-linked cascade in Ras-transformed cells. *Free Radical Biology and Medicine, 44*(4), 624–634. https://doi.org/10.1016/j.freeradbiomed.2007.10.041.

Choy, M. S., Li, Y., Machado, L., Kunze, M. B. A., Connors, C. R., Wei, X., ... Peti, W. (2017). Conformational rigidity and protein dynamics at distinct timescales regulate PTP1B activity and allostery. *Molecular Cell, 65*(4), 644–658.e645. https://doi.org/10.1016/j.molcel.2017.01.014.

Cohen, P. (2001). The role of protein phosphorylation in human health and disease. The Sir Hans Krebs Medal Lecture. *European Journal of Biochemistry, 268*(19), 5001–5010. https://doi.org/10.1046/j.0014-2956.2001.02473.x.

Crean, R. M., Biler, M., van der Kamp, M. W., Hengge, A. C., & Kamerlin, S. C. L. (2021). Loop dynamics and enzyme catalysis in protein tyrosine phosphatases. *Journal of the American Chemical Society, 143*(10), 3830–3845. https://doi.org/10.1021/jacs.0c11806.

Cui, D. S., Lipchock, J. M., Brookner, D., & Loria, J. P. (2019). Uncovering the molecular interactions in the catalytic loop that modulate the conformational dynamics in protein tyrosine phosphatase 1B. *Journal of the American Chemical Society, 141*(32), 12634–12647. https://doi.org/10.1021/jacs.9b04470.

Cunnick, J. M., Dorsey, J. F., Mei, L., & Wu, J. (1998). Reversible regulation of SHP-1 tyrosine phosphatase activity by oxidation. *IUBMB Life, 45*(5), 887–894. https://doi.org/10.1002/iub.7510450506.

Dagnell, M., Cheng, Q., & Arnér, E. S. J. (2021). Qualitative differences in protection of PTP1B activity by the reductive Trx1 or TRP14 enzyme systems upon oxidative challenges with polysulfides or H(2)O(2) together with bicarbonate. *Antioxidants (Basel), 10*(1), https://doi.org/10.3390/antiox10010111.

Dagnell, M., Cheng, Q., Rizvi, S. H. M., Pace, P. E., Boivin, B., Winterbourn, C. C., & Arnér, E. S. J. (2019). Bicarbonate is essential for protein-tyrosine phosphatase 1B (PTP1B) oxidation and cellular signaling through EGF-triggered phosphorylation cascades. *Journal of Biological Chemistry, 294*(33), 12330–12338. https://doi.org/10.1074/jbc.RA119.009001.

Dagnell, M., Frijhoff, J., Pader, I., Augsten, M., Boivin, B., Xu, J., ... Östman, A. (2013). Selective activation of oxidized PTP1B by the thioredoxin system modulates PDGF-β receptor tyrosine kinase signaling. *Proceedings of the National Academy of Sciences, 110*(33), 13398–13403. https://doi.org/10.1073/pnas.1302891110.

Denu, J. M., & Dixon, J. E. (1998). Protein tyrosine phosphatases: Mechanisms of catalysis and regulation. *Current Opinion in Chemical Biology, 2*, 633–641. https://doi.org/10.1016/S1367-5931(98)80095-1.

Denu, J. M., Lohse, D. L., Vijayalakshmi, J., Saper, M. A., & Dixon, J. E. (1996). Visualization of intermediate and transition-state structures in protein-tyrosine phosphatase catalysis. *Proceedings of the National Academy of Sciences of the United States of America, 93*(6), 2493–2498. Retrieved from. http://www.ncbi.nlm.nih.gov/entrez/query.fcgi?cmd=Retrieve&db=PubMed&dopt=Citation&list_uids=8637902.

Di Marzo, N., Chisci, E., & Giovannoni, R. (2018). The role of hydrogen peroxide in redox-dependent signaling: homeostatic and pathological responses in mammalian cells. *Cells, 7*(10), https://doi.org/10.3390/cells7100156.

Dokainish, H. M., & Gauld, J. W. (2015). Formation of a stable iminol intermediate in the redox regulation mechanism of protein tyrosine phosphatase 1B (PTP1B). *ACS Catalysis, 5*(4), 2195–2202. https://doi.org/10.1021/cs501707h.

Eswaran, J., von Kries, J. P., Marsden, B., Longman, E., Debreczeni, J. E., Ugochukwu, E., ... Barr, A. J. (2006). Crystal structures and inhibitor identification for PTPN5, PTPRR and PTPN7: A family of human MAPK-specific protein tyrosine phosphatases. *Biochemical Journal, 395*(3), 483–491. https://doi.org/10.1042/BJ20051931.

Frijhoff, J., Dagnell, M., Godfrey, R., & Östman, A. (2013). Regulation of protein tyrosine phosphatase oxidation in cell adhesion and migration. *Antioxidants & Redox Signaling, 20*(13), 1994–2010. https://doi.org/10.1089/ars.2013.5643.

Guan, K. L., & Dixon, J. E. (1991). Evidence for protein-tyrosine-phosphatase catalysis proceeding via a cysteine-phosphate intermediate. *Journal of Biological Chemistry, 266*(26), 17026–17030. Retrieved from http://www.ncbi.nlm.nih.gov/entrez/query.fcgi?cmd=Retrieve&db=PubMed&dopt=Citation&list_uids=1654322.

Hansen, S. K., Cancilla, M. T., Shiau, T. P., Kung, J., Chen, T., & Erlanson, D. A. (2005). Allosteric inhibition of PTP1B activity by selective modification of a non-active site cysteine residue. *Biochemistry, 44*(21), 7704–7712. https://doi.org/10.1021/bi047417s.

Haque, A., Andersen, J. N., Salmeen, A., Barford, D., & Tonks, N. K. (2011). Conformation-sensing antibodies stabilize the oxidized form of PTP1B and inhibit its phosphatase activity. *Cell, 147*(1), 185–198. https://doi.org/10.1016/j.cell.2011.08.036.

Hay, I. M., Fearnley, G. W., Rios, P., Köhn, M., Sharpe, H. J., & Deane, J. E. (2020). The receptor PTPRU is a redox sensitive pseudophosphatase. *Nature Communications, 11*(1), 3219. https://doi.org/10.1038/s41467-020-17076-w.

Heo, S., Kim, S., & Kang, D. (2020). The role of hydrogen peroxide and peroxiredoxins throughout the cell cycle. *Antioxidants (Basel), 9*(4), https://doi.org/10.3390/antiox9040280.

Hunter, T. (1995). Protein kinases and phosphatases: the yin and yang of protein phosphorylation and signaling. *Cell, 80*(2), 225–236. 0092-8674(95)90405-0.

Iwamoto, N., Sumi, D., Ishii, T., Uchida, K., Cho, A. K., Froines, J. R., & Kumagai, Y. (2007). Chemical knockdown of protein-tyrosine phosphatase 1B by 1,2-naphthoquinone through covalent modification causes persistent transactivation of epidermal growth factor receptor. *Journal of Biological Chemistry, 282*(46), 33396–33404. https://doi.org/10.1074/jbc.M705224200.

Jeon, T. J., Chien, P. N., Chun, H.-J., & Ryu, S. E. (2013). Structure of the catalytic domain of protein tyrosine phosphatase sigma in the sulfenic acid form. *Molecular Cell, 36*(1), 55–61. https://doi.org/10.1007/s10059-013-0033-x.

Juarez, J. C., Manuia, M., Burnett, M. E., Betancourt, O., Boivin, B., Shaw, D. E., ... Doñate, F. (2008). Superoxide dismutase 1 (SOD1) is essential for H2O2-mediated oxidation and inactivation of phosphatases in growth factor signaling. *Proceedings of the National Academy of Sciences of the United States of America, 105*(20), 7147–7152. https://doi.org/10.1073/pnas.0709451105.

Khan, S., Bjij, I., & Soliman, M. E. S. (2019). Selective covalent inhibition of "Allosteric Cys121" distort the binding of PTP1B enzyme: A novel therapeutic approach for cancer treatment. *Cell Biochemistry and Biophysics, 77*(3), 203–211. https://doi.org/10.1007/s12013-019-00882-5.

Krishnan, N., Bonham, C. A., Rus, I. A., Shrestha, O. K., Gauss, C. M., Haque, A., ... Tonks, N. K. (2018). Harnessing insulin- and leptin-induced oxidation of PTP1B for therapeutic development. *Nature Communications, 9*(1), 283. https://doi.org/10.1038/s41467-017-02252-2.

Lennicke, C., Rahn, J., Lichtenfels, R., Wessjohann, L. A., & Seliger, B. (2015). Hydrogen peroxide – production, fate and role in redox signaling of tumor cells. *Cell Communication and Signaling, 13*(1), 39. https://doi.org/10.1186/s12964-015-0118-6.

Londhe, A. D., Bergeron, A., Curley, S. M., Zhang, F., Rivera, K. D., Kannan, A., ... Boivin, B. (2020). Regulation of PTP1B activation through disruption of redox-complex formation. *Nature Chemical Biology, 16*(2), 122–125. https://doi.org/10.1038/s41589-019-0433-0.

Lou, Y.-W., Chen, Y.-Y., Hsu, S.-F., Chen, R.-K., Lee, C.-L., Khoo, K.-H., ... Meng, T.-C. (2008). Redox regulation of the protein tyrosine phosphatase PTP1B in cancer cells. *The FEBS Journal, 275*(1), 69–88. https://doi.org/10.1111/j.1742-4658.2007.06173.x.

Machado, L. E. S. F., Critton, D. A., Page, R., & Peti, W. (2017). Redox regulation of a gain-of-function mutation (N308D) in SHP2 Noonan syndrome. *ACS Omega, 2*(11), 8313–8318. https://doi.org/10.1021/acsomega.7b01318.

Machado, L. E. S. F., Shen, T.-L., Page, R., & Peti, W. (2017). The KIM-family protein-tyrosine phosphatases use distinct reversible oxidation intermediates: Intramolecular or intermolecular disulfide bond formation. *Journal of Biological Chemistry, 292*(21), 8786–8796. https://doi.org/10.1074/jbc.M116.774174.

Madan, L. L., & Gopal, B. (2011). Conformational basis for substrate recruitment in protein tyrosine phosphatase 10D. *Biochemistry, 50*(46), 10114–10125. https://doi.org/10.1021/bi201092q.

Madan, L. L., Goutam, K., & Gopal, B. (2012). Inter-domain interactions influence the stability and catalytic activity of the bi-domain protein tyrosine phosphatase PTP99A. *Biochimica et Biophysica Acta, 1824*(8), 983–990. https://doi.org/10.1016/j.bbapap.2012.05.001.

Madan, L. L., Veeranna, S., Shameer, K., Reddy, C. C. S., Sowdhamini, R., & Gopal, B. (2011). Modulation of catalytic activity in Multi-Domain protein tyrosine phosphatases. *PLoS One, 6.* https://doi.org/10.1371/journal.pone.0024766.

Mahadev, K., Zilbering, A., Zhu, L., & Goldstein, B. J. (2001). Insulin-stimulated hydrogen peroxide reversibly inhibits protein-tyrosine phosphatase 1b in vivo and enhances the early insulin action cascade. *Journal of Biological Chemistry, 276*(24), 21938–21942. https://doi.org/10.1074/jbc.C100109200.

Mailloux, R. J. (2015). Teaching the fundamentals of electron transfer reactions in mitochondria and the production and detection of reactive oxygen species. *Redox Biology, 4,* 381–398. https://doi.org/10.1016/j.redox.2015.02.001.

Manning, G., Whyte, D. B., Martinez, R., Hunter, T., & Sudarsanam, S. (2002). The protein kinase complement of the human genome. *Science, 298*(5600), 1912–1934. https://doi.org/10.1126/science.1075762.

Marinho, H. S., Real, C., Cyrne, L., Soares, H., & Antunes, F. (2014). Hydrogen peroxide sensing, signaling and regulation of transcription factors. *Redox Biology, 2,* 535–562. https://doi.org/10.1016/j.redox.2014.02.006.

Mesquita, F. S., Dyer, S. N., Heinrich, D. A., Bulun, S. E., Marsh, E. E., & Nowak, R. A. (2010). Reactive oxygen species mediate mitogenic growth factor signaling pathways in human leiomyoma smooth muscle cells. *Biology of Reproduction, 82*(2), 341–351. https://doi.org/10.1095/biolreprod.108.075887.

Moise, G., Morales, Y., Beaumont, V., Caradonna, T., Loria, J. P., Johnson, S. J., & Hengge, A. C. (2018). A YopH PTP1B chimera shows the importance of the WPD-loop sequence to the activity, structure, and dynamics of protein tyrosine phosphatases. *Biochemistry, 57*(36), 5315–5326. https://doi.org/10.1021/acs.biochem.8b00663.

Muise, E. S., Vrielink, A., Ennis, M. A., Lemieux, N. H., & Tremblay, M. L. (1996). Thermosensitive mutants of the MPTP and hPTP1B protein tyrosine phosphatases: Isolation and structural analysis. *Protein Science, 5,* 604–613. https://doi.org/10.1002/pro.5560050405.

Netto, L. E., & Antunes, F. (2016). The roles of peroxiredoxin and thioredoxin in hydrogen peroxide sensing and in signal transduction. *Molecular Cell, 39*(1), 65–71. https://doi.org/10.14348/molcells.2016.2349.

Netto, L. E. S., & Machado, L. E. S. F. (2022). Preferential redox regulation of cysteine-based protein tyrosine phosphatases: Structural and biochemical diversity. *The FEBS Journal, 289*(18), 5480–5504. https://doi.org/10.1111/febs.16466.

Ohba, M., Shibanuma, M., Kuroki, T., & Nose, K. (1994). Production of hydrogen peroxide by transforming growth factor-beta 1 and its involvement in induction of egr-1 in mouse osteoblastic cells. *Journal of Cell Biology, 126*(4), 1079–1088. https://doi.org/10.1083/jcb.126.4.1079.

Ostman, A., Frijhoff, J., Sandin, A., & Böhmer, F. D. (2011). Regulation of protein tyrosine phosphatases by reversible oxidation. *Journal of Biochemistry, 150*(4), 345–356. https://doi.org/10.1093/jb/mvr104.

Palde, P. B., & Carroll, K. S. (2015). A universal entropy-driven mechanism for thioredoxin–target recognition. *Proceedings of the National Academy of Sciences, 112*(26), 7960–7965. https://doi.org/10.1073/pnas.1504376112.

Perkins, A., Nelson, K. J., Parsonage, D., Poole, L. B., & Karplus, P. A. (2015). Peroxiredoxins: Guardians against oxidative stress and modulators of peroxide signaling. *Trends in Biochemical Sciences, 40*(8), 435–445. https://doi.org/10.1016/j.tibs.2015.05.001.

Peskin, A. V., Low, F. M., Paton, L. N., Maghzal, G. J., Hampton, M. B., & Winterbourn, C. C. (2007). The high reactivity of peroxiredoxin 2 with $H(2)O(2)$ is not reflected in its reaction with other oxidants and thiol reagents. *Journal of Biological Chemistry, 282*(16), 11885–11892. https://doi.org/10.1074/jbc.M700339200.

Peti, W., & Page, R. (2015). Strategies to make protein serine/threonine (PP1, calcineurin) and tyrosine phosphatases (PTP1B) druggable: Achieving specificity by targeting substrate and regulatory protein interaction sites. *Bioorganic & Medicinal Chemistry, 23*, 2781–2785. https://doi.org/10.1016/j.bmc.2015.02.040.

Prasad, S., Gupta, S. C., & Tyagi, A. K. (2017). Reactive oxygen species (ROS) and cancer: Role of antioxidative nutraceuticals. *Cancer Letters, 387*, 95–105. https://doi.org/10.1016/j.canlet.2016.03.042.

Psenakova, K., Hexnerova, R., Srb, P., Obsilova, V., Veverka, V., & Obsil, T. (2020). The redox-active site of thioredoxin is directly involved in apoptosis signal-regulating kinase 1 binding that is modulated by oxidative stress. *The FEBS Journal, 287*(8), 1626–1644. https://doi.org/10.1111/febs.15101.

Punthasee, P., Laciak, A. R., Cummings, A. H., Ruddraraju, K. V., Lewis, S. M., Hillebrand, R., ... Gates, K. S. (2017). Covalent allosteric inactivation of protein tyrosine phosphatase 1B (PTP1B) by an inhibitor-electrophile conjugate. *Biochemistry, 56*(14), 2051–2060. https://doi.org/10.1021/acs.biochem.7b00151.

Ravichandran, L. V., Chen, H., Li, Y., & Quon, M. J. (2001). Phosphorylation of PTP1B at Ser(50) by Akt impairs its ability to dephosphorylate the insulin receptor. *Molecular Endocrinology, 15*(10), 1768–1780. https://doi.org/10.1210/mend.15.10.0711.

Reczek, C. R., & Chandel, N. S. (2015). ROS-dependent signal transduction. *Current Opinion in Cell Biology, 33*, 8–13. https://doi.org/10.1016/j.ceb.2014.09.010.

Rhee, S. G., & Woo, H. A. (2011). Multiple functions of peroxiredoxins: Peroxidases, sensors and regulators of the intracellular messenger H_2O_2, and protein chaperones. *Antioxidants & Redox Signaling, 15*(3), 781–794. https://doi.org/10.1089/ars.2010.3393.

Salmeen, A., Andersen, J. N., Myers, M. P., Meng, T.-C., Hinks, J. A., Tonks, N. K., & Barford, D. (2003). Redox regulation of protein tyrosine phosphatase 1B involves a sulphenyl-amide intermediate. *Nature, 423*, 769–773. https://doi.org/10.1038/nature01680.

Scapin, G., Patel, S., Patel, V., Kennedy, B., & Asante-Appiah, E. (2001). The structure of apo protein-tyrosine phosphatase 1B C215S mutant: More than just an S − > O change. *Protein Science, 10*(8), 1596–1605. https://doi.org/10.1110/ps.11001.

Sies, H. (2017). Hydrogen peroxide as a central redox signaling molecule in physiological oxidative stress: Oxidative eustress. *Redox Biology, 11*, 613–619. https://doi.org/10.1016/j.redox.2016.12.035.

Sundaresan, M., Yu, Z. X., Ferrans, V. J., Irani, K., & Finkel, T. (1995). Requirement for generation of H2O2 for platelet-derived growth factor signal transduction. *Science, 270*(5234), 296–299. https://doi.org/10.1126/science.270.5234.296.

Svegliati, S., Cancello, R., Sambo, P., Luchetti, M., Paroncini, P., Orlandini, G., ... Gabrielli, A. (2005). Platelet-derived Growth Factor and Reactive Oxygen Species (ROS) Regulate Ras Protein Levels in Primary Human Fibroblasts via $ERK_{1/2}$: AMPLIFICATION OF ROS AND Ras IN SYSTEMIC SCLEROSIS FIBROBLASTS. *Journal of Biological Chemistry, 280*(43), 36474–36482. https://doi.org/10.1074/jbc.M502851200.

Tanner, J. J., Parsons, Z. D., Cummings, A. H., Zhou, H., & Gates, K. S. (2011). Redox regulation of protein tyrosine phosphatases: Structural and chemical aspects. *Antioxidants & Redox Signaling, 15*(1), 77–97. https://doi.org/10.1089/ars.2010.3611.

Toledano-Katchalski, H., Tiran, Z., Sines, T., Shani, G., Granot-Attas, S., Den Hertog, J., & Elson, A. (2003). Dimerization in vivo and inhibition of the nonreceptor form of protein tyrosine phosphatase epsilon. *Molecular and Cellular Biology, 23*(15), 5460–5471. https://doi.org/10.1128/mcb.23.15.5460-5471.2003.

Tonks, N. K. (2003). PTP1B: From the sidelines to the front lines!. *FEBS Letters, 546*(1), 140–148. https://doi.org/10.1016/S0014-5793(03)00603-3.

Tonks, N. K. (2005). Redox redux: revisiting PTPs and the control of cell signaling. *Cell, 121*(5), 667–670. https://doi.org/10.1016/j.cell.2005.05.016.

Tonks, N. K. (2006). Protein tyrosine phosphatases: From genes, to function, to disease. *Nature Reviews. Molecular Cell Biology, 7*, 833–846. https://doi.org/10.1038/nrm2039.

Tonks, N. K. (2013). Protein tyrosine phosphatases—From housekeeping enzymes to master regulators of signal transduction. *The FEBS Journal, 280*(2), 346–378. https://doi.org/10.1111/febs.12077.

Tonks, N. K., Diltz, C. D., & Fischer, E. H. (1988). Purification of the major protein-tyrosine-phosphatases of human placenta. *Journal of Biological Chemistry, 263*(14), 6722–6730. Retrieved from. https://www.ncbi.nlm.nih.gov/pubmed/2834386.

Torgeson, K. R., Clarkson, M. W., Kumar, G. S., Page, R., & Peti, W. (2020). Cooperative dynamics across distinct structural elements regulate PTP1B activity. *Journal of Biological Chemistry, 295*(40), 13829–13837. https://doi.org/10.1074/jbc.RA120.014652.

Tsai, S. J., Sen, U., Zhao, L., Greenleaf, W. B., Dasgupta, J., Fiorillo, E., ... Chen, X. S. (2009). Crystal structure of the human lymphoid tyrosine phosphatase catalytic domain: Insights into redox regulation. *Biochemistry, 48*(22), 4838–4845. https://doi.org/10.1021/bi900166y.

Tsutsumi, R., Harizanova, J., Stockert, R., Schröder, K., Bastiaens, P. I. H., & Neel, B. G. (2017). Assay to visualize specific protein oxidation reveals spatio-temporal regulation of SHP2. *Nature Communications, 8*(1), 466. https://doi.org/10.1038/s41467-017-00503-w.

van Montfort, R. L. M., Congreve, M., Tisi, D., Carr, R., & Jhoti, H. (2003). Oxidation state of the active-site cysteine in protein tyrosine phosphatase 1B. *Nature, 423*(6941), 773–777. https://doi.org/10.1038/nature01681.

Vilchis-Landeros, M. M., Matuz-Mares, D., & Vázquez-Meza, H. (2020). Regulation of metabolic processes by hydrogen peroxide generated by NADPH oxidases. *Processes, 8*(11), https://doi.org/10.3390/pr8111424.

Weibrecht, I., Böhmer, S.-A., Dagnell, M., Kappert, K., Östman, A., & Böhmer, F.-D. (2007). Oxidation sensitivity of the catalytic cysteine of the protein-tyrosine phosphatases SHP-1 and SHP-2. *Free Radical Biology and Medicine, 43*(1), 100–110. https://doi.org/10.1016/j.freeradbiomed.2007.03.021.

Welsh, C. L., Allen, S., & Madan, L. K. (2023). Setting sail: Maneuvering SHP2 activity and its effects in cancer. *Advances in Cancer Research, 160*, 17–60. https://doi.org/10.1016/bs.acr.2023.03.003.

Welsh, C. L., Pandey, P., & Ahuja, L. G. (2021). Chapter Seven – Protein tyrosine phosphatases: A new paradigm in an old signaling system? In K. D. Tew, & P. B. Fisher (Vol. Eds.), *Advances in Cancer Research: Vol. 152*, (pp. 263–303). Academic Press.

Weng, M. S., Chang, J. H., Hung, W. Y., Yang, Y. C., & Chien, M. H. (2018). The interplay of reactive oxygen species and the epidermal growth factor receptor in tumor progression and drug resistance. *Journal of Experimental & Clinical Cancer Research, 37*(1), 61. https://doi.org/10.1186/s13046-018-0728-0.

Whittier, S. K., Hengge, A. C., & Loria, J. P. (2013). Conformational motions regulate phosphoryl transfer in related protein tyrosine phosphatases. *Science, 341*(6148), 899–903. https://doi.org/10.1126/science.1241735.

Winterbourn, C. C., & Hampton, M. B. (2008). Thiol chemistry and specificity in redox signaling. *Free Radical Biology and Medicine, 45*(5), 549–561. https://doi.org/10.1016/j.freeradbiomed.2008.05.004.

Winterbourn, C. C., & Metodiewa, D. (1999). Reactivity of biologically important thiol compounds with superoxide and hydrogen peroxide. *Free Radical Biology and Medicine, 27*(3), 322–328. https://doi.org/10.1016/S0891-5849(99)00051-9.

Yarnall, M. T. N., Kim, S. H., Korntner, S., & Bishop, A. C. (2022). Destabilization of the SHP2 and SHP1 protein tyrosine phosphatase domains by a non-conserved "backdoor" cysteine. *Biochemistry and Biophysics Reports, 32*, 101370. https://doi.org/10.1016/j.bbrep.2022.101370.

Zhang, Z.-Y. (2017). Drugging the undruggable: Therapeutic potential of targeting protein tyrosine phosphatases. *Accounts of Chemical Research, 50*(1), 122–129. https://doi.org/10.1021/acs.accounts.6b00537.

Zhang, Z.-Y., Wang, Y., Wu, L., Fauman, E. B., Stuckey, J. A., Schubert, H. L., ... Dixon, J. E. (1994). The Cys(X)5Arg catalytic motif in phosphoester hydrolysis. *Biochemistry, 33*, 15266–15270. https://doi.org/10.1021/bi00255a007.

Zhang, Z. Y., & Dixon, J. E. (1993). Active site labeling of the Yersinia protein tyrosine phosphatase: The determination of the pKa of the active site cysteine and the function of the conserved histidine 402. *Biochemistry, 32*(36), 9340–9345. https://doi.org/10.1021/bi00087a012.

Zhang, Z. Y., Wang, Y., & Dixon, J. E. (1994). Dissecting the catalytic mechanism of protein-tyrosine phosphatases. *Proceedings of the National Academy of Sciences of the United States of America, 91*, 1624–1627. https://doi.org/10.1073/pnas.91.5.1624.

Zhao, Y., Wu, L., Noh, S. J., Guan, K. L., & Zhang, Z. Y. (1998). Altering the nucleophile specificity of a protein-tyrosine phosphatase-catalyzed reaction. Probing the function of the invariant glutamine residues. *Journal of Biological Chemistry, 273*(10), 5484–5492. https://doi.org/10.1074/jbc.273.10.5484.

CHAPTER THREE

Mitochondrial metallopeptidase OMA1 in cancer

Gunjan Purohit[a], Polash Ghosh[a], and Oleh Khalimonchuk[a,b,c,*]

[a]Department of Biochemistry, University of Nebraska-Lincoln, Lincoln, NE, United States
[b]Nebraska Redox Biology Center, Lincoln, NE, United States
[c]Fred & Pamela Buffett Cancer Center, Omaha, NE, United States
*Corresponding author. e-mail address: okhalimonchuk2@unl.edu

Contents

1. Introduction	75
2. OMA1 and its role in mitochondrial and cellular physiology	78
3. Redox regulation of OMA1	82
4. OMA1 and its regulation in cancers	83
5. Clinical implications	89
6. Challenges and future directions	90
Acknowledgments	91
Conflict of interest	91
References	91

Abstract

Our understanding of the roles that mitochondria play in cellular physiology has evolved drastically—from a mere cellular energy supplier to a crucial regulator of metabolic and signaling processes, particularly in the context of development and progression of human diseases such as cancers. The present review examines the role of OMA1, a conserved, redox-sensitive metallopeptidase in cancer biology. OMA1's involvement in mitochondrial quality control, redox activity, and stress responses underscores its potential as a novel target in cancer diagnosis and treatment. However, our incomplete understanding of OMA1's regulation and structural detail presents ongoing challenges to target OMA1 for therapeutic purposes. Further exploration of OMA1 holds promise in uncovering novel insights into cancer mechanisms and therapeutic strategies. In this chapter, we briefly summarize our current knowledge about OMA1, its redox-regulation, and emerging role in certain cancers.

1. Introduction

Historically, mitochondria were primarily considered as the energy hubs of cells, but mounting evidence highlights their central role in a plethora of cellular processes and life-threatening ailments such as cancer. There

Advances in Cancer Research, Volume 162
ISSN 0065-230X, https://doi.org/10.1016/bs.acr.2024.05.001
Copyright © 2024 Elsevier Inc. All rights are reserved, including those for text and data mining, AI training, and similar technologies.

is a growing understanding of the mitochondrial components and processes implicated in cancer development and progression. Mitochondrial functions—including bioenergetics, metabolism, and signaling—are now widely recognized as crucial mediators of tumor growth and metastasis.

One of the most common features of cancer cells is their altered metabolism, wherein they exhibit a shift towards glycolysis despite the presence of normal amounts of oxygen. This phenomenon—discovered almost a century ago and known as the "Warburg effect"—underscores metabolic alterations and the heightened energy demand of rapidly proliferating cancer cells (Warburg, 1956). Although our interpretation of this phenomenon has evolved dramatically, it remains an important hallmark of many cancers. Indeed, many types of cancer demonstrate normal mitochondrial oxidative phosphorylation (OXPHOS) activity alongside intensified aerobic glycolysis, facilitating the production of necessary metabolites and building molecules essential for cancer growth, a state known as anaplerosis (Ahn & Metallo, 2015; Jose, Bellance, & Rossignol, 2011; Moreno-Sánchez et al., 2014). Moreover, cancer cells induce metabolic changes in surrounding immune cells—to evade identification, destruction, and support tumorigenesis (Guerra, Bonetti, & Brenner, 2020; Leone & Powell, 2021). Growing evidence suggests that altered mitochondrial function supports cancer cells across various stages of tumorigenesis, including survival, proliferation, and metastasis (Gonzalez, Hagerling, & Werb, 2018).

In most cells, mitochondria are organized into highly interconnected tubular networks. As such, one emerging controller of the aforementioned alterations is mitochondrial dynamics—a process defined as changes in mitochondrial network's shape, number, and position within a cell. This behavior is mediated by two reciprocal events known as fission and fusion of the mitochondrial network, which occur in response to various physiological stimuli. Mitochondrial dynamics is central to cellular physiology and plays a critical role in human pathologies including oncological disorders by influencing cellular metabolism, Ca^{2+} ion fluxes, proliferative and differentiation properties, autophagy flux, and apoptotic resistance (Chan, 2020; Chen, Zhao, & Li, 2023). Dysregulation of mitochondrial dynamics also constitutes a set of well-recognized molecular hallmarks in various cancers, thus suggesting that altered mitochondrial behavior may also contribute to cancer-related mechanisms. For example, aberrant mitochondrial dynamics has been reported in glioma, neuroblastoma, lung carcinoma, and colon cancer (Ma, Wang, & Jia, 2020; Maycotte et al., 2017). Likewise, altered expression of mitochondrial behavior-mediating factors has been reported in certain cancers, including invasive breast carcinoma (Boulton & Caino, 2022; Xing et al., 2022). Moreover, genetic or pharmacological inhibition of mitochondrial network partitioning results in reduced proliferation and tumor growth in several types of cancer (Rodrigues & Ferraz, 2020). On the other hand, enhanced mitochondrial fusion is known to exert anti-apoptotic and pro-survival effects and has been postulated to contribute to a mechanism of chemoresistance in neuroblastoma and gynecological cancers (Casinelli et al., 2016; Genovese et al., 2021; Kong et al., 2015; Suen, Norris, & Youle, 2008).

The fusion and partitioning of the mitochondrial network involve both the outer (OM) and inner (IM) mitochondrial membranes, each of which is a coordinated, yet physically separate process. At present, it is believed that molecular events in the IM appear to precede ones that involve the OM (Adebayo, Singh, Singh, & Dasgupta, 2022; Kleele et al., 2021). Of note, most of the previous research has focused primarily on the OM aspect of mitochondrial dynamics in cancer growth. For example, OM fission–mediating guanosine triphosphatase (GTPase) dynamin related protein1 (DRP1) was shown to be an important downstream effector of KRAS–dependent oncogenic signaling, which occurs through ERK1/2–mediated fission–promoting phosphorylation of DRP1 (Kashatus et al., 2016; Ponsoda et al., 2016). Additionally, ERK kinase regulates the activity of pro-fusion OM GTPase MFN1, thereby also stimulating fission (Pyakurel et al., 2015). Similarly, the c–MYC oncoprotein impacts mitochondrial behavior via modulation of the expression of OM pro-fission factors, including DRP1, as well as signaling pathways downstream of mitochondrial dynamics in breast cancer (Kortlever et al., 2015). The fusion and fission of the IM as well as its pro-apoptotic remodeling are mediated by OPA1 GTPase and are far less understood (Frezza et al., 2006; Ishihara, Fujita, Oka, & Mihara, 2006; Song, Chen, Fiket, Alexander, & Chan, 2007).

In the present chapter, we will primarily focus on the IM protease OMA1 (Overlapping Activity with m-AAA Protease 1), its redox regulation, and role in cancer biology. OMA1 plays a pivotal role in regulating mitochondrial dynamics—primarily through stress-processing of OPA1 GTPase and subsequent alteration of mitochondrial behavior. It is also involved in quality control of OXPHOS-related proteins and—most recently—has been implicated in the regulation of integrated stress response (ISR) signaling. This occurs via OMA1-mediated processing of DELE1 protein, leading to cytoplasmic accumulation of the short form of DELE1 that primes the signaling axis encompassing heme-regulated inhibitor (HRI) kinase and Eukaryotic

Initiation Factor 2 alpha (eIF2α) kinase, leading in turn to ISR-mediated induction of stress-associated Activating Transcription Factor 4 (ATF4) that orchestrates adaptive expression of certain genes, thereby promoting cell survival (Fessler et al., 2020; Guo et al., 2020). Therefore, OMA1 impacts cancer-related processes by influencing mitochondrial behavior, bioenergetics, metabolism, and stress responses. As such, deeper understanding of the mechanisms and implications of OMA1 in cancer could offer insights into this important regulator of mitochondrial physiology and related therapeutic targets or strategies to modulate mitochondrial function in cancer cells. The OMA1's central role in mitochondrial physiology that extends far beyond classical quality control functions makes it an intriguing protein, and its relevance to cancer remains an active area of investigation. Further research into the connections between OMA1 and cancer holds the promise of unveiling novel insights into the underlying mechanisms of tumorigenesis and potential new therapeutic avenues for cancer treatment.

2. OMA1 and its role in mitochondrial and cellular physiology

OMA1 is the IM-localized conserved enzyme that belongs to the family of zinc metallopeptidases or metzincins (Gomis-Ruth, 2009; López-Pelegrín et al., 2013). OMA1 plays a critical role in maintaining mitochondrial quality control and homeostasis, particularly in response to homeostatic challenges, including metabolic fluctuations and changes in the organelle's redox state. Anchored in the IM, OMA1 interacts with various mitochondrial proteins in this sub-compartment. The C-terminal region of OMA1 protrudes into the intermembrane space (IMS) and harbors the enzyme's catalytic domain (Käser, Kambacheld, Kisters-Woike, & Langer, 2003). While largely dormant under the basal conditions, OMA1 undergoes conformational changes, thereby becoming rapidly activated through yet incompletely understood mechanism (s). In so doing, it dramatically reshapes mitochondrial dynamics by processing and cleaving another IM-resident factor—Optic Atrophy 1 (OPA1)—the conserved GTPase that is known to play a crucial role in mitochondrial fusion and division, cristae remodeling, and cell death initiation (Ehses et al., 2009; Head, Griparic, Amiri, Gandre-Babbe, & Van Der Bliek, 2009) (Fig. 1). Mounting evidence indicates that stress-activated OMA1 actively engages in other critical mitochondrial processes such as stress signaling, which is orchestrated in a similar proteolytic fashion.

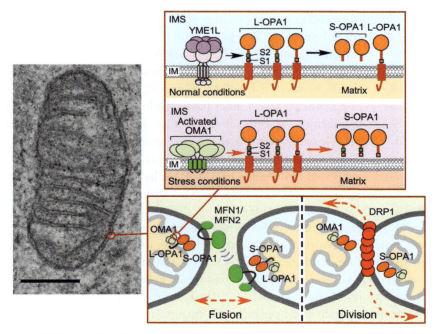

Fig. 1 **OMA1-mediated mitochondrial membrane and network remodeling.** The schematics show principal mechanisms through which mitochondrial membranes are remodeled in response to various stimuli. Bottom panel summarizes the overall process of mitochondrial network fusion and division and key factors mediating mitochondrial membrane dynamics. Some additional factors are not depicted. The top right panel focuses on the process of the inner mitochondrial membrane dynamics, which is mediated through selective processing of the long OPA1 GTPase (L-OPA1) isoform to its short variant (S-OPA1) by IM-anchored protease YME1L and stress-activated metallopeptidase OMA1. A representative transmission electron microscopy micrograph of mitochondrion from wild-type mouse embryonic fibroblasts shown on the left is used to highlight the site where the events in question are taking place. *IM*, Inner mitochondrial membrane; *IMS*, intermembrane space; *L-OPA1/S-OPA1*, long and short forms of the IM GTPase OPA1, respectively. Bar, 0.4 mm.

At present, the OMA1-mediated processing of OPA1 is perhaps the protease's best understood function through which it controls the mitochondrial network morphology, consequently adjusting organelle's bioenergetic, metabolic, and signaling outputs and their capacity to adapt to varying physiological demands. Indeed, by teaming up with another conserved IM protease called YME1L or i-AAA (for IMS-facing AAA protease), OMA1 mediates functional integrity of the inner membrane when cellular homeostasis is challenged by a variety of stimuli, including uncoupling, oligomerization of pro-apoptotic proteins in the OM, redox insults, or nutrient limitation. The pro-fusion GTPase OPA1 exists in two variants: a long, IM-anchored form (L-OPA1) and a short form peripheral to the IM (S-OPA1), which under basal conditions is produced by YME1L-mediated cleavage of L-OPA1 (Griparic, Kanazawa, & Van Der Bliek, 2007; Song et al., 2007). Under basal conditions, both variants exist roughly in equilibrium, as only about half of L-OPA1 splice variants contain YME1L cleavage sites. At the same time, all L-OPA1 forms are equipped with the OMA1 cleavage site known as S1. Homeostatic insults trigger stress-activated processing of L-OPA1 by OMA1, thereby converting all long variants of the GTPase into S-OPA1 species and promoting IM fission (Anand et al., 2014; Head et al., 2009; Zhang, Li, & Song, 2014) (Fig. 1). In turn, this propels downstream events, such as partitioning of the mitochondrial network, initiation of mitophagy, or apoptosis (Baker et al., 2014). Therefore, stabilization of L-OPA1 through OMA1 depletion leads to marked apoptotic resistance and reportedly is beneficial in post-mitotic cells and tissues (Acin-Perez et al., 2018; Korwitz et al., 2016). However, in other models such as cancer cells, this manipulation has been reported to have adverse effects (Quirós et al., 2012). Moreover, OMA1 activity is required for the balanced and tuneable respiratory function in yeast and mammalian cells, thereby assuring optimal bioenergetic output during homeostatic and metabolic challenges (Bohovych, Donaldson, Christianson, Zahayko, & Khalimonchuk, 2014, 2015, 2016; Bohovych, Dietz, Swenson, Zahayko, & Khalimonchuk, 2019; Quirós et al., 2012; Viana, Levytskyy, Anand, Reichert, & Khalimonchuk, 2021).

Recent high-throughput genetic screens identified another key substrate of OMA1 called DELE1 (Fessler et al., 2020; Guo et al., 2020). The DAP3 Binding Cell Death Enhancer 1, or DELE1—a scaffold-like protein that oligomerizes and primes the formation of the heme-regulated factor HRI-eIF2α kinases signaling axis that dovetails with the survival-promoting integrated stress response pathway (Fessler et al., 2020; Guo et al., 2020). Under normal physiological conditions, DELE1 in its presumably inactive form is at least partially imported into the mitochondria where it faces the IMS. Homeostatic insults activate OMA1, which subsequently cleaves DELE1, generating a shortened form of the protein (S-DELE1), whereupon it slips back into the cytoplasm, binds and activates HRI, subsequently leading to eIF2α phosphorylation and, consequently, ISR (Guo et al., 2020). As pointed out earlier, the ISR plays a crucial role in supporting cancer cell survival under homeostatically challenging conditions of high metabolic demand and tumor microenvironment conditions by promoting the synthesis of essential proteins needed for rapid cell growth (Winter, Yadav, & Rutter, 2022) (Fig. 2). Moreover, ISR activation in cancer cells can also contribute towards resistance against chemotherapy and targeted therapy regimens (Borankova et al., 2023). Therefore, targeting the ISR pathway is increasingly recognized as a promising therapeutic strategy in cancer treatment.

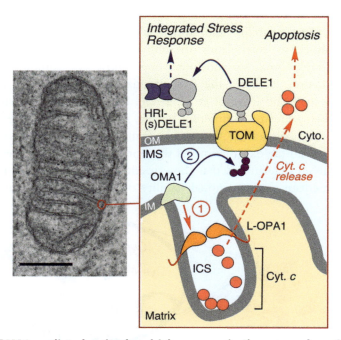

Fig. 2 OMA1-mediated mitochondrial communication cascades. Schematic depiction of two currently known OMA1-orchestrated signaling pathways that link mitochondrial stress to global cellular death and pro-survival responses. (1) Stress-activated OMA1 cleaves the L-OPA1 isoform, which in addition to its role in the IM dynamics is also crucial for mitochondrial cristae remodeling. Consequently, this process can facilitate cytochrome c release from mitochondria, thereby priming canonical apoptotic cascade. (2) Stress-triggered OMA1 cleaves DELE1, thereby yielding a truncated variant of the protein—designated (s)DELE1—that escapes to the cytosol where it oligomerizes and sequentially recruits HRI and eIF2α kinases. Subsequent eIF2α phosphorylation triggers a multifaceted integrated stress response program, leading to cancer cell survival. The crosstalk and functional cooperation between said pathways as well as further molecular details thereof remain to be elucidated.

3. Redox regulation of OMA1

Studies by our laboratory and others have established that OMA1 exists in a semi-reduced state and its activity can be modulated by reducing or oxidizing conditions, whereby the enzyme's activity decreases under reducing conditions (Bohovych et al., 2019; Miallot et al., 2023). Studies in baker's yeast model identified two highly conserved cysteine residues, Cys-272 and Cys-332, and showed that disulfide bonds formed by these residues play an important and conserved role in the regulation of Oma1 function (Bohovych et al., 2019). Intriguingly, Cys-332 is an invariant residue that is critical for Oma1 stability and even substituting this amino acid with residues mimicking sulfenylation or sulfhydryl sidechains of cysteine did not yield a stable Oma1 variant. Interestingly, however, combined C272A C332A mutation stabilizes the mutant enzyme, underscoring the significance of the bond (Miallot et al., 2023). Structural modeling indicates that while being located in different helices, these two residues reside in a close proximity, suggesting they could form an intramolecular disulfide (Fig. 3), although the exact nature of this bond remains to be investigated. Further analyses determined that these residues play a structural role and appear to mediate conformational stability of the Oma1 complex. Thus,

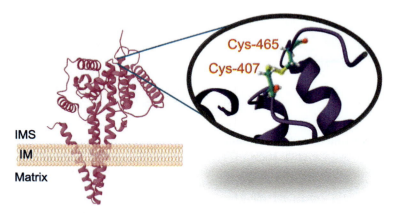

Fig. 3 Molecular organization of OMA1 metallopeptidase. The cartoon shows the topology of the AlphaFold-predicted mature (381 amino acids) human OMA1 metallopeptidase. The enzyme is anchored in the inner mitochondrial membrane (IM) with its proteolytic domain facing the intermembrane space (IMS). The protein's unstructured N-terminal region (residues 1–143) harboring mitochondrial targeting sequence is removed for clarity. The inset shows conserved Cys-407 and Cys-465 residues (full-length protein numbering) that form a disulfide bond critical for the enzyme function. *IM*, inner mitochondrial membrane; *IMS*, intermembrane space.

changes in the redox state of mitochondria can regulate Oma1 stability and activity through Cys-272 and Cys-332. Moreover, studies with a chimeric model substrate of Oma1—called S1–1 and encompassing a fragment harboring the enzyme's naturally occurring cleavage site—showed that Oma1 interacts with the substrate in a redox-sensitive manner (Bohovych et al., 2019). The follow-up work by another laboratory has demonstrated that a corresponding set of cysteine residues in mammalian OMA1—the Cys-403 in particular—acts in a similar redox-sensing switch manner in the rodent enzyme (Miallot et al., 2023). This work has been carried out in a context of *in vivo* mouse sarcoma model underscoring pathophysiological significance of said redox sensing switch in cancer development. Indeed, mutating Cys-403 to alanine resulted in altered physiological responses to stress and enhanced mitochondrial DNA release, thereby promoting anti-tumor immunity and preventing tumor development in immunocompetent animals. Further studies are warranted to better understand the molecular mechanisms behind these observations.

4. OMA1 and its regulation in cancers

In recent years, multiple studies implicated OMA1 in cancer development or progression and identified the protease as prospective diagnostic or therapeutic target in various types of cancer. Historically, several high-throughput studies first have identified *OMA1* as an increased chromosomal instability gene, a loss of heterozygosity site in genetic models of cancer (Srinivasainagendra et al., 2017; Yuen et al., 2007), and one of the top candidate homozygous deletion genes in colorectal tumors (Leary et al., 2008). These findings, however, remained largely unnoticed until later as OMA1 was not a specific focus of said studies.

An important insight regarding OMA1's role in breast cancer was gained through an initial observation that OMA1-deficient mouse embryonic fibroblasts (MEF) exhibit increased aerobic glycolysis and enhanced proliferative properties upon culturing conditions that require maximal bioenergetic output (Daverey et al., 2019; Viana et al., 2021). Furthermore, OMA1-deficient MEF cells were found to exhibit metabolic signatures indicative of a shift towards anaplerotic metabolism. In line with this notion, OMA1-deleted MEF cells were more reliant on glutaminolysis, i.e., displayed enhanced reductive carboxylation of glutamine to yield glutamate, which is subsequently modified to α-ketoglutarate and utilized in Krebs cycle

(Bohovych et al., 2015). Consequently, OMA1-deleted MEF cells were more sensitive to inhibition of glutaminase—a key enzyme that mediates glutamine to glutamate conversion (Daverey et al., 2019). Subsequent studies with OMA1-depleted 21MT-1 patient-derived metastatic breast cancer cells validated these findings and established that OMA1 suppresses proliferation and migratory properties of metastatic breast adenocarcinoma cells. Conversely, OMA1 depletion in 21MT-1 cells augmented their invasiveness by inducing epithelial to mesenchymal transition (Daverey et al., 2019). Of interest, a comparative analysis of OMA1 expression levels and overall survival of breast cancer patients using data available through The Cancer Genome Atlas (TCGA) database identified a strong correlation between low OMA1 expression and poor prognosis in patients afflicted with breast cancer (Daverey et al., 2019) (Fig. 4). Furthermore, TCGA, Molecular Taxonomy of Breast Cancer International Consortium (METABRIC), and Human Protein Atlas Consortium data indicate that OMA1 levels are lower in the most aggressive triple-negative breast cancer (TNBC; negative for estrogen receptor, progesterone receptor, and ErbB2 receptor) tissues as compared to normal and luminal breast cancer tissues. These observations suggest that activation of OMA1 could be a potential therapeutic or adjuvant strategy in TNBC treatment. Intriguingly, findings from the Human Protein Atlas Consortium indicate extremely low OMA1 protein levels not only in breast cancer tissues, but also in testicular cancer tissues and certain lymphomas, suggesting possible OMA1 inactivation in these cancers (Uhlén et al., 2015).

The role of OMA1 in certain types of cancer appears to be more stochastic and tissue context-dependent. A recent study has shown that in contrast to breast cancer, hypoxic colorectal cancer cells display OMA1-dependent metabolic reprogramming, supporting cancer cell growth by promoting glycolysis and inhibiting oxidation phosphorylation functions (Wu et al., 2021). Likewise, the loss of OMA1 has been shown to have an anti-tumorigenic effect in azoxymethane/dextran sodium sulfate-induced model of inflammatory colorectal cancer (Wu et al., 2021). Similarly, in gastric cancer patients, high OMA1 expression is associated with decreased expression of mitochondrial outer membrane protein mitofusin MFN2 and predicts unfavorable prognosis (Amini et al., 2020; Zhang et al., 2013). Furthermore, OMA1 has been identified as a target of miR-433-3p microRNA in osteosarcoma (Li et al., 2023). The upregulation of long non-coding RNA PCGEM1 protects OMA1 from degradation by miR-433-3p, promoting osteosarcoma growth (Li et al., 2023). Conversely, knockdown of PCGEM1 suppresses osteosarcoma progression through miR-433-3p mediated degradation of

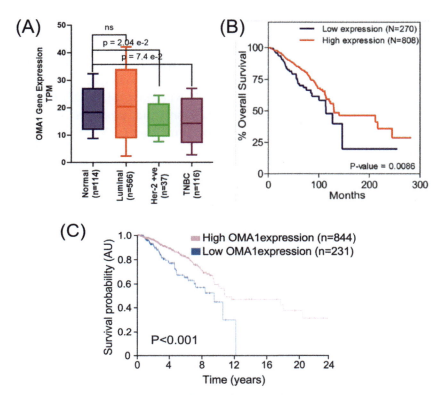

Fig. 4 Association between the levels of OMA1 expression and survival of patients with breast cancer. (A). Box plot showing OMA1 mRNA expression from normal, luminal, HER2-positive, and triple-negative (negative for estrogen, progesterone, and ErbB2 receptors, TNBC) breast cancer samples from TCGA Breast Adenocarcinoma Dataset. Boxes show minimum and maximum—bottom and top lines, respectively-and mean (line within the boxes) values. A paired t-test was used to calculate statistical significance. TPM, transcripts per kilobase of exon per million mapped reads; ns, not significant. (B). Kaplan-Meier curve shows overall survival percentage with patients breast tumors expressing low or high OMA1 from TCGA Breast Adenocarcinoma Dataset. P-value was calculated using a log-rank test. (C). Survival data comparison OMA1-High and OMA1-Low protein data from Protein Atlas Dataset also indicates that breast cancer patients with low OMA1 protein levels exhibited significantly poor survival when compared to patients with high OMA1 protein levels. P-value calculated using a log-rank test.

OMA1, whereas this effect is reversed by OMA1 overexpression. A recent study provided a possible explanation for such seemingly contradicting behavior of OMA1 in different types of cancer. Rivera–Mejias et al. determined that OMA1 acts as a metabolic safeguard in control of cellular glucose metabolism, thereby ensuring tissue/environment context-specific cell adaptations and survival under severe cellular stress conditions such as nuclear DNA damage and consequent p53-dependent apoptosis (Rivera-Mejías et al., 2023). Related to this notion, another study reported on important metabolic role of OMA1 in the context of heart failure mouse model, wherein OMA1-DELE1-HRI signaling axis supports glutathione metabolism and promotes selenium-dependent accumulation of the glutathione peroxidase GPX4, thereby protecting cardiomyocytes from ferroptotic cell death (Acin-Perez et al., 2018).

While future studies are warranted to better understand the exact molecular underpinnings of the above effects of OMA1 in various types of cancer, several non-mutually exclusive candidate mechanisms are currently considered. One scenario involves the OMA1-DELE1-HRI signaling cascade whereby stress-activated OMA1 activates HRI-eIF2α signaling axis of ISR via cleavage of DELE1, leading to ATF4 induction and promoting cancer cell survival (Fessler et al., 2020; Guo et al., 2020) (Fig. 2). Consistent with this model, phosphorylated eIF2α is significantly upregulated in breast cancer patients and predicts disease-free survival in individuals afflicted with triple negative breast cancer (Guo et al., 2017).

The second scenario pertains to a relatively well-understood role of OMA1 in the regulation of mitochondrial network dynamics. As mentioned earlier, OMA1 undergoes rapid activation by a variety of cellular insults, including changes in metabolic demands or homeostatic challenges (Ehses et al., 2009; Korwitz et al., 2016). Once activated, OMA1 executes proteolytic conversion of L-OPA1 to its short form, thus promoting cristae remodeling, IM fission, and subsequent fragmentation of the mitochondrial network—the events required for downstream processes such as mitochondrial morphology, changes in metabolic activity, apoptosis, and mitophagy (Anand et al., 2014; Baker et al., 2014; Bohovych, Donaldson, Christianson, Zahayko, & Khalimonchuk, 2014; Ishihara, Fujita, Oka, & Mihara, 2006; MacVicar & Lane, 2014; Quirós et al., 2012; Rainbolt, Lebeau, Puchades, & Wiseman, 2016; Varanita et al., 2015; Zhang, Li, & Song, 2014). Prevention of L-OPA1 processing via OMA1 attenuation in cancer cells can stabilize the IM ultrastructure and mitochondrial network, thus inhibiting apoptotic activity and promoting pro-survival effects in cancer cells (Anand et al., 2014; MacVicar & Lane, 2014; Quirós et al., 2012; Varanita et al., 2015). Indeed, alterations in fusion or fission dynamics have been linked to changes in nutrient utilization and energy conversion in various tumor cells (Akita et al., 2014; Hagenbuchner, Kuznetsov, Obexer, & Ausserlechner, 2013; Kashatus et al., 2016; Kashatus, 2019; Qian et al., 2014; Roy, Reddy, Iijima, & Sesaki, 2015; Serasinghe et al., 2015; Wai & Langer, 2016;

Westermann, 2012; Zhao et al., 2013). Similarly, enhanced mitochondrial fragmentation observed in different cancer cells correlates with glycolytic metabolism and not oxidative phosphorylation. Furthermore, KRAS oncogenic signaling that is known to induce glycolytic switching in tumors also promotes mitochondrial division (Kashatus et al., 2016; Serasinghe et al., 2015), which almost inevitably involves the OMA1-OPA1 cascade. Moreover, such changes in mitochondrial network behavior have been linked to regulation of cell "stemness", a cell's propensity to self-renew/differentiate. Mitochondrial division mediated through ERK1/2 kinases and pro-fission factor, dynamin related protein DRP1, were shown to underlie reprograming of somatic adult cells into induced pluripotent stem cells (Prieto et al., 2016). On the other hand, profusion activity of OPA1 and mitofusins appears to be required for maintenance of stem cell identity and self-renewal capacity in several types of stem cells (Luchsinger, de Almeida, Corrigan, Mumau, & Snoeck, 2016; Nassar & Blanpain, 2016; Patten et al., 2014). Of note, many tumors harbor sub-populations of cells that also exhibit hallmarks of "stemness" and appear to determine malignant growth and metastatic activity of the tumor (Luchsinger et al., 2016; Nassar & Blanpain, 2016; Prieto et al., 2016). In line with this notion, enhanced DRP1-mediated fission dynamics have been reported in brain tumorigenic cells (Xie et al., 2015), thereby underscoring the role of mitochondrial network dynamics in the fate of tumor initiating cells. It is noteworthy that mitochondrial behavior is intricately linked to the cell cycle (Chang & Blackstone, 2010; Lee, Westrate, Wu, Page, & Voeltz, 2016). For instance, DRP1 activity is regulated by the Cyclin B-CDK1 kinase cell cycle-controlling axis as well as SUMO-specific peptidase SenP5 during mitosis (Lee et al., 2016; Xie, Wang, Jiang, & Li, 2019; Zunino, Braschi, Xu, & McBride, 2009). Reciprocally, inhibition of DRP1-mediated mitochondrial fragmentation can delay mitotic progression and—consequently—cellular proliferation (Xie et al., 2019; Zunino et al., 2009). Cell cycle-regulating APC/C E3 ubiquitin ligase also promotes tubulation of the mitochondrial network during G1/S phase transition through attenuation of DRP1 (Horn et al., 2011; Mitra, Wunder, Roysam, Lin, & Lippincott-Schwartz, 2009). In line with this notion, c-MYC oncogene-induced changes in mitochondrial network dynamics in breast cells were shown to limit YAP/TAZ signaling, thereby promoting tumor growth (Kortlever et al., 2015) OMA1 deficit is known to be sufficient to overpower DRP1-mediated cues (MacVicar and Lane, 2014), thereby making the protease a potential key contributor to tumor pathophysiology.

It is noteworthy that OMA1 abundance is also regulated at a post-translational level. In addition to cleaving its cognizant substrates such as

OPA1 and DELE1, stress-activated OMA1 undergoes autoproteolysis (Baker et al., 2014; Head et al., 2009; Zhang et al., 2014), which likely represents a control feedback mechanism to remove the activated enzyme and minimize potential collateral damage. This also underscores the fact that both transcript and protein levels of OMA1 should be evaluated in cancer cells or samples of interest to reach definitive conclusions when correlating OMA1 expression with prognosis in certain cancers. Related to this notion, a functional crosstalk between OMA1 and another mitochondrial protease, YME1L—also known as i-AAA protease—has been identified. Briefly, OMA1 unrestrained activity has been shown to be one of the main pathological drivers in cardiac YME1L-deleted mice. Conversely, deletion of OMA1 in the animals lacking cardiac YME1L mitigated any pathological manifestations (Wai et al., 2015). In agreement, studies in human neuronal cultured cells have shown that YME1L can supress OMA1 activity and vice versa, depending on the energy/ nutrient status of the organelle—e.g., in ATP-depleted or depolarized mitochondria (Rainbolt et al., 2016). It remains to be clarified, however, whether such a reciprocity exists in other cell types, which were suggested to rely primarily on OMA1's autoproteolysis rather than its removal via YME1L as a feedback loop mechanism. These findings are significant, as YME1L itself has been shown to mediate communication of cellular nutrient status to mitochondria via mTORC1-Lipin cascade, leading to decreased phosphatidylethanolamine levels in the IM and subsequent YME1L activation and changes in mitochondrial and cellular metabolism, which appears to play a role in cancers such as pancreatic ductal adenocarcinoma (PDAC) (Mac Vicar et al., 2019). Considering the aforementioned functional interconnectedness of YME1L and OMA1, it is likely that the latter could be also involved in said regulation. Of note, in addition to PDAC, YME1L deregulation has been observed in colorectal cancer, glioma, and ovarian cancer (Cao et al., 2023; Liao, Chu, Su, Wu, & Li, 2022; Mac Vicar et al., 2019; Srinivasainagendra et al., 2017). In addition to YME1L, other factors have been reported to stimulate OMA1 turnover under specific conditions (Anderson et al., 2020; Rainbolt et al., 2016; Yang et al., 2018). For instance, prohibitins PHB1 and PHB2 were found to exert pro-tumorigenic effects in neuroblastoma, in part by stabilizing L-OPA1 in a cardiolipin-dependent manner (Anderson et al., 2020). Conversely, reduced prohibitin levels trigger OMA1-mediated integrated stress response and cell death in neuroblastoma (Borankova et al., 2023). In ovarian cancer, PHB1 was reported to enhance OMA1-mediate processing of L-OPA1. Furthermore, PHB1 directly interacts with phosphorylated p53 and BAK, thereby correlating with better prognosis and chemoresponsiveness (Kong, Wang, Fung, Xue, & Tsang, 2014). Future studies are needed to better understand these effects and explore their therapeutic potential.

5. Clinical implications

Adaptive alterations in key mitochondrial pathways that underlie metabolic rewiring present potential opportunities to develop new therapeutic strategies in cancer treatment. Our knowledge of OMA1 as an emerging master controller in mitochondrial physiology and metabolism is expanding and so does the recognition of its value as a prospective therapeutic target in various cancers. For instance, cisplatin—a common chemotherapeutic agent, elicits genotoxic effects and oxidative stress and induces mitochondrial fragmentation in chemosensitive gynecological cancers, but not in chemoresistant cancer cells (Kong et al., 2014). In chemosensitive cancer cells, cisplatin treatment activates OMA1 and increases L-OPA1 processing—leading to mitochondrial fragmentation and cell death due to cristae remodeling and release of cytochrome c, which is necessary to initiate canonical apoptosis. However, in chemoresistant cells this process often appears to be compromised (Kong et al., 2014). In line with this notion are the observations that cells overexpressing L-OPA1 or depleted of OMA1 exhibit marked apoptotic resistance (Varanita et al., 2015; Viana et al., 2021). Similarly, knockdown of p53 attenuates cisplatin-induced activation of OMA1 in chemosensitive cancer cells (Kong et al., 2014). By contrast, overexpression of p53 in chemoresistant p53-null ovarian cancer cells induces OMA1-mediated L-OPA1 processing, making them more chemosensitive. Furthermore, piceatannol, a plant metabolite, has been found to potentiate cisplatin sensitivity in chemoresistant ovarian cancer cells (Farrand et al., 2013). The effect appears to be mediated by p53 and leads to increased expression of the pro-apoptotic protein NOXA and consequent enhanced caspase-3 activation. Considering that p53 regulates OMA1 activity in ovarian cancer and the protease's role in apoptotic priming, clarifying the role of OMA1 in piceatannol-adjuvant treatment of ovarian cancer cells may hold clues to new therapeutic strategies.

Related to the above discussion, the OMA1-OPA1 functional axis has been proposed to contribute to resistance of acute myeloid leukemia cells to the targeted therapy drug venetoclax (Chen et al., 2019). Clearly, more research is warranted to understand the mechanism in question.

Very recently, a novel class of compounds has been identified in a search of new drugs to treat aggressive B-cell lymphoma (Schwarzer, Oliveira, & Kleppa, 2023). Modified pyrazolo-thiazole compounds called BTM-3528 and BTM-3566, were found to be effective against diffuse large B cell lymphoma (DLBCL) in preclinical models. As a single agent, these compounds were able to achieve 30%–36% responses in first-line chemotherapy and about 20% responses in the second line. Mechanistically, in DLBCL cells these compounds activate OMA1, which subsequently induces DELE1-dependent ISR and L-OPA1 cleavage, ultimately resulting in extensive fragmentation of mitochondrial network and apoptosis (Schwarzer et al., 2023). This study also identified a poorly characterized IM factor called FAM210B as a negative regulator of BTM-3528 and BTM-3566 activity. Overexpression of FAM210B prevented the compounds-induced OMA1 activation and apoptosis, although the molecular bases of this effect remain to be further investigated.

In chemoresistant pediatric T-cell acute lymphoblastic leukemia (T-ALL), chemosensitivity can be achieved by elevating reactive oxygen species (ROS) production (Silic-Benussi et al., 2018). Treating T-ALL cells with a K^+ channel activator NS1619 while simultaneously inhibiting the pentose phosphate pathway by dehydroepiandrosterone (DHEA) has been shown to reduce ROS scavenging (Silic-Benussi et al., 2018). Importantly, treatment with NS1619 and DHEA stress-activated the OMA1-OPA1 axis in T-ALL cells, whereas silencing of OMA1-inhibited L-OPA1 processing leading to cell death. Considering that OMA1 depletion inhibits canonical apoptosis, it will be interesting to investigate which type of cell death occurs in this case.

Taken together, these findings indicate that targeting OMA1—either solely or in combination with other drugs or treatment regimen—in various cancers can be a promising therapeutic strategy in cancer.

6. Challenges and future directions

While mounting evidence underscores translational potential and therapeutic promise of OMA1 protease and molecular pathways that it orchestrates in cancer, a number of challenges remain. First, the structural details of OMA1 and mechanistic basis of its stress sensing and activation remain unresolved, which obscures our understanding of its activity and limits the use of structural information for guided design of specific inhibitors or activators. Second, limited information is available on genomic regulation of OMA1. No systematic analysis has been carried out to acquire and validate information about the OMA1 promoter, regulatory sequences, relevant transcription and enhancer factors, or histone modifications in normal and disease states, including details about alternative and/or tissue-specific transcripts and splice variants. Similarly, the process of post-translational maturation and modification of the OMA1 protein is incompletely understood and requires further studies. Finally, a more thorough mechanistic understanding of the pathways through which OMA1 exerts its regulatory effects on mitochondrial network behavior, cellular physiology, signaling and immunometabolic cues is much needed to gain fundamental understanding of this fascinating enzyme and unlock its diagnostic and therapeutic potential in different types of cancer.

Acknowledgments

We apologize to those authors whose work we were unable to cite due to space constraints. We wish to thank the members of the Khalimonchuk laboratory for helpful discussions and editorial help and Drs. Vimla Band and Asher Rajan for their help with cancer patients database mining. We also wish to acknowledge the grant support from the National Institutes of Health, GM131701–01 to O.K.

Conflict of interest

The authors declare that they have no conflicts of interest with the contents of this text.

References

Acin-Perez, R., Lechuga-Vieco, A. V., DelMarMuñoz, M., Nieto-Arellano, R., Torroja, C., Sánchez-Cabo, F., ... Enríquez, J. A. (2018). Ablation of the stress protease OMA1 protects against heart failure in mice. *Science Translational Medicine, 10*. https://doi.org/10.1126/scitranslmed.aan4935.

Adebayo, M., Singh, S., Singh, A. P., & Dasgupta, S. (2022). Mitochondrial fusion and fission: The fine-tune balance for cellular homeostasis. *The FASEB Journal, 35*, 1–23. https://doi.org/10.1096/fj.202100067R.

Ahn, C. S., & Metallo, C. M. (2015). Mitochondria as biosynthetic factories for cancer proliferation. *Cancer & Metabolism, 3*, 1–10. https://doi.org/10.1186/s40170-015-0128-2.

Akita, M., Suzuki-Karasaki, M., Fujiwara, K., Nakagawa, C., Soma, M., Yoshida, Y., ... Suzuki-Karasaki, Y. (2014). Mitochondrial division inhibitor-1 induces mitochondrial hyperfusion and sensitizes human cancer cells to TRAIL-induced apoptosis. *International Journal of Oncology, 45*, 1901–1912. https://doi.org/10.3892/ijo.2014.2608.

Amini, M. A., Karimi, J., Khodadadi, I., Tavilani, H., Talebi, S. S., & Afshar, B. (2020). Overexpression of ROMO1 and OMA1 are potentially biomarkers and predict unfavorable prognosis in gastric cancer. *Journal of Gastrointestinal Cancer, 51*, 939–946. https://doi.org/10.1007/s12029-019-00330-w.

Anand, R., Wai, T., Baker, M. J., Kladt, N., Schauss, A. C., Rugarli, E., & Langer, T. (2014). The i-AAA protease YME1L and OMA1 cleave OPA1 to balance mitochondrial fusion and fission. *The Journal of Cell Biology, 204*, 919–929. https://doi.org/10.1083/jcb.201308006.

Anderson, C. J., Kahl, A., Fruitman, H., Qian, L., Zhou, P., Manfredi, G., & Iadecola, C. (2020). Prohibitin levels regulate OMA1 activity and turnover in neurons. *Cell Death and Differentiation, 27*, 1896–1906. https://doi.org/10.1038/s41418-019-0469-4.

Baker, M. J., Lampe, P. A., Stojanovski, D., Korwitz, A., Anand, R., Tatsuta, T., & Langer, T. (2014). Stress-induced OMA1 activation and autocatalytic turnover regulate OPA1-dependent mitochondrial dynamics. *The EMBO Journal, 33*, 578–593. https://doi.org/10.1002/embj.201386474.

Bohovych, I., Donaldson, G., Christianson, S., Zahayko, N., & Khalimonchuk, O. (2014). Stress-triggered activation of the metalloprotease Oma1 involves its C-terminal region and is important for mitochondrial stress protection in yeast. *The Journal of Biological Chemistry, 289*, 13259–13272. https://doi.org/10.1074/jbc.M113.542910.

Bohovych, I., Fernandez, M. R., Rahn, J. J., Stackley, K. D., Bestman, J. E., Anandhan, A., ... Khalimonchuk, O. (2015). Metalloprotease OMA1 fine-tunes mitochondrial bioenergetic function and respiratory supercomplex stability. *Scientific Reports, 5*(1), 14. https://doi.org/10.1038/srep13989.

Bohovych, I., Kastora, S., Christianson, S., Topil, D., Kim, H., Fangman, T., ... Khalimonchuk, O. (2016). Oma1 links mitochondrial protein quality control and TOR signaling to modulate physiological plasticity and cellular stress responses. *Molecular and Cellular Biology, 36*, 2300–2312. https://doi.org/10.1128/mcb.00156-16.

Bohovych, I., Dietz, J. V., Swenson, S., Zahayko, N., & Khalimonchuk, O. (2019). Redox regulation of the mitochondrial quality control protease Oma1. *Antioxidants Redox Signal, 31*, 429–443. https://doi.org/10.1089/ars.2018.7642.

Borankova, K., Krchniakova, M., Leck, L. Y. W., Neradil, J., Jansson, P. J., Hogarty, M. D., & Skoda, J. (2023). Inhibiting mitochondrial translation overcomes multidrug resistance in MYC-driven neuroblastoma via OMA1-mediated integrated stress response. *BioRxiv*, 2023.02.24. https://doi.org/10.1101/2023.02.24.529852.

Boulton, D. P., & Caino, M. C. (2022). Mitochondrial fission and fusion in tumor progression to metastasis. *Frontiers in Cell and Developmental Biology, 10*, 1–19. https://doi.org/10.3389/fcell.2022.849962.

Cao, C., Liu, F., Chen, G., Zhou, L.-N., Wang, Y., Zhang, Z.-Q., & Qin, X. (2023). YME1L overexpression exerts pro-tumorigenic activity in glioma by promoting G α i1 expression and Akt activation. *Protein & Cells, 14*, 223–229.

Casinelli, G., Larosa, J., Sharma, M., Cherok, E., Banerjee, S., Branca, M., ... Graves, J. A. (2016). N-Myc overexpression increases cisplatin resistance in neuroblastoma via deregulation of mitochondrial dynamics. *Cell Death Discovery, 2*, 1–11. https://doi.org/10.1038/cddiscovery.2016.82.

Chan, D. C. (2020). Mitochondrial dynamics and its involvement in disease. *Annual Review of Pathology, 15*, 235–259. https://doi.org/10.1146/annurev-pathmechdis-012419-032711.

Chang, C.-R., & Blackstone, C. (2010). Dynamic regulation of mitochondrial fission through modification of the dynamin-related protein Drp1. *Annals of the New York Academy of Sciences, 1201*, 34–39. https://doi.org/10.1111/j.1749-6632.2010.05629.x.

Chen, X., Glytsou, C., Zhou, H., Narang, S., Reyna, D. E., Lopez, A., ... Aifantis, I. (2019). Targeting mitochondrial structure sensitizes acute myeloid leukemia to venetoclax treatment. *Cancer Discovery, 9*, 890–909. https://doi.org/10.1158/2159-8290.CD-19-0117.

Chen, W., Zhao, H., & Li, Y. (2023). Mitochondrial dynamics in health and disease: Mechanisms and potential targets. *Signal Transduction and Targeted Therapy, 8*. https://doi.org/10.1038/s41392-023-01547-9.

Daverey, A., Levytskyy, R. M., Stanke, K. M., Viana, M. P., Swenson, S., Hayward, S. L., ... Kidambi, S. (2019). Depletion of mitochondrial protease OMA1 alters proliferative properties and promotes metastatic growth of breast cancer cells. *Scientific Reports, 9,* 1–15. https://doi.org/10.1038/s41598-019-49327-2.

Ehses, S., Raschke, I., Mancuso, G., Bernacchia, A., Geimer, S., Tondera, D., ... Langer, T. (2009). Regulation of OPA1 processing and mitochondrial fusion by m-AAA protease isoenzymes and OMA1. *The Journal of Cell Biology, 187,* 1023–1036. https://doi.org/10.1083/jcb.200906084.

Farrand, L., Byun, S., Kim, J. Y., Im-aram, A., Lee, J., Lim, S., ... Tsang, B. K. (2013). Piceatannol enhances cisplatin sensitivity in ovarian cancer via modulation of p53, X-linked inhibitor of apoptosis protein (XIAP), and mitochondrial fission. *Journal of Biological Chemistry, 288,* 23740–23750. https://doi.org/10.1074/jbc.M113.487686.

Fessler, E., Eckl, E. M., Schmitt, S., Mancilla, I. A., Meyer-Bender, M. F., Hanf, M., ... Jae, L. T. (2020). A pathway coordinated by DELE1 relays mitochondrial stress to the cytosol. *Nature, 579,* 433–437. https://doi.org/10.1038/s41586-020-2076-4.

Frezza, C., Cipolat, S., Martins de Brito, O., Micaroni, M., Beznoussenko, G. V., Rudka, T., ... Scorrano, L. (2006). OPA1 controls apoptotic cristae remodeling independently from mitochondrial fusion. *Cell, 126,* 177–189. https://doi.org/10.1016/j.cell.2006.06.025.

Genovese, I., Carinci, M., Modesti, L., Aguiari, G., Pinton, P., & Giorgi, C. (2021). Mitochondria: insights into crucial features to overcome cancer chemoresistance. *International Journal of Molecular Sciences, 22,* 1–30.

Gomis-Ruth, F. X. (2009). Catalytic domain architecture of metzincin metalloproteases. *The Journal of Biological Chemistry, 284,* 15353–15357. https://doi.org/10.1074/jbc.R800069200.

Gonzalez, H., Hagerling, C., & Werb, Z. (2018). Roles of the immune system in cancer: From tumor initiation to metastatic progression. *Genes & Development, 32,* 1267–1284. https://doi.org/10.1101/gad.314617.118.

Griparic, L., Kanazawa, T., & Van Der Bliek, A. M. (2007). Regulation of the mitochondrial dynamin-like protein Opa1 by proteolytic cleavage. *JCB, 178,* 757–764. https://doi.org/10.1083/jcb.200704112.

Guerra, L., Bonetti, L., & Brenner, D. (2020). Metabolic modulation of immunity: A new concept in cancer immunotherapy. *Cell Reports, 32,* 107848. https://doi.org/10.1016/j.celrep.2020.107848.

Guo, L., Chi, Y., Xue, J., Ma, L., Shao, Z., & Wu, J. (2017). Phosphorylated eIF2 α predicts disease-free survival in triple- negative breast cancer patients. *Scientific Reports, 7,* 1–9. https://doi.org/10.1038/srep44674.

Guo, X., Aviles, G., Liu, Y., Tian, R., Unger, B. A., Lin, Y. H. T., ... Kampmann, M. (2020). Mitochondrial stress is relayed to the cytosol by an OMA1–DELE1–HRI pathway. *Nature, 579,* 427–432. https://doi.org/10.1038/s41586-020-2078-2.

Hagenbuchner, J., Kuznetsov, A. V., Obexer, P., & Ausserlechner, M. J. (2013). BIRC5/ Survivin enhances aerobic glycolysis and drug resistance by altered regulation of the mitochondrial fusion/fission machinery. *Oncogene, 32,* 4748–4757. https://doi.org/10.1038/onc.2012.500.

Head, B., Griparic, L., Amiri, M., Gandre-Babbe, S., & Van Der Bliek, A. M. (2009). Inducible proteolytic inactivation of OPA1 mediated by the OMA1 protease in mammalian cells. *The Journal of Cell Biology, 187,* 959–966. https://doi.org/10.1083/jcb.200906083.

Horn, S. R., Thomenius, M. J., Johnson, E. S., Freel, C. D., Wu, J. Q., Coloff, J. L., ... Kornbluth, S. (2011). Regulation of mitochondrial morphology by APC/CCdh1-mediated control of Drp1 stability. *Molecular Biology of the Cell, 22,* 1207–1216. https://doi.org/10.1091/mbc.E10-07-0567.

Ishihara, N., Fujita, Y., Oka, T., & Mihara, K. (2006). Regulation of mitochondrial morphology through proteolytic cleavage of OPA1. *The EMBO Journal, 25*, 2966–2977. https://doi.org/10.1038/sj.emboj.7601184.

Jose, C., Bellance, N., & Rossignol, R. (2011). Choosing between glycolysis and oxidative phosphorylation: A tumor's dilemma? *Biochimica et Biophysica Acta (BBA) – Bioenergetics, 1807*, 552–561. https://doi.org/10.1016/j.bbabio.2010.10.012.

Käser, M., Kambacheld, M., Kisters-Woike, B., & Langer, T. (2003). Oma1, a novel membrane-bound metallopeptidase in mitochondria with activities overlapping with the m-AAA protease. *The Journal of Biological Chemistry, 278*, 46414–46423. https://doi.org/10.1074/jbc.M305584200.

Kashatus, J. A., Nascimento, A., Myers, L. J., Sher, A., Frances, L., Hoehn, K. L., ... Kashatus, D. F. (2016). Erk2 phosphorylation of Drp1 promotes mitochondrial fission and MAPK-driven tumor growth. *Molecular Cell, 57*, 537–551. https://doi.org/10.1016/j.molcel.2015.01.002.

Kashatus, D. F. (2019). The regulation of tumor cell physiology by mitochondrial dynamics. *Biochemical and Biophysical Research Communications, 27*, 9–16. https://doi.org/10.1016/j.bbrc.2017.06.192.

Kleele, T., Rey, T., Winter, J., Zaganelli, S., Mahecic, D., Lambert, H. P., ... Mahecic, D. (2021). Distinct fission signatures predict mitochondrial degradation or biogenesis. *Nature, 593*, 435–439. https://doi.org/10.1038/s41586-021-03510-6.

Kong, B., Wang, Q., Fung, E., Xue, K., & Tsang, B. K. (2014). P53 is required for cis-platin-induced processing of the mitochondrial fusion protein L-Opa1 that is mediated by the mitochondrial metallopeptidase Oma1 in gynecologic cancers. *The Journal of Biological Chemistry, 289*, 27134–27145. https://doi.org/10.1074/jbc.M114.594812.

Kong, B., Tsuyoshi, H., Orisaka, M., Shieh, D.-B., Yoshida, Y., & Tsang, B. K. (2015). Mitochondrial dynamics regulating chemoresistance in gynecological cancers. *Annals of the New York Academy of Sciences, 1350*, 1–16. https://doi.org/10.1111/nyas.12883.

Kortlever, R. M., Evan, G. I., Kempa, S., Eilers, M., Eyss, V., Jaenicke, L. A., ... Letschert, S. (2015). A MYC-driven change in mitochondrial dynamics limits YAP/TAZ function in mammary epithelial cells and breast cancer. *Cancer Cell, 28*, 743–757. https://doi.org/10.1016/j.ccell.2015.10.013.

Korwitz, A., Merkwirth, C., Richter-Dennerlein, R., Tröder, S. E., Sprenger, H. G., Quirós, P. M., ... Langer, T. (2016). Loss of OMA1 delays neurodegeneration by preventing stress-induced OPA1 processing in mitochondria. *The Journal of Cell Biology, 212*, 157–166. https://doi.org/10.1083/jcb.201507022.

Leary, R. J., Lin, J. C., Cummins, J., Boca, S., Wood, L. D., Parsons, D. W., ... Velculescu, V. E. (2008). Integrated analysis of homozygous deletions, focal amplifications, and sequence alterations in breast and colorectal cancers. *Proceedings of the National Academy of Sciences of the United States of America, 105*, 16224–16229. https://doi.org/10.1073/pnas.0808041105.

Lee, J. E., Westrate, L. M., Wu, H., Page, C., & Voeltz, G. K. (2016). Multiple dynamin family members collaborate to drive mitochondrial division. *Nature, 540*, 139–143. https://doi.org/10.1038/nature20555.

Leone, R. D., & Powell, J. D. (2021). Metabolism of immune cells in cancer. *Nature Reviews Cancer, 20*, 516–531. https://doi.org/10.1038/s41568-020-0273-y.

Li, J., Zhang, Y., Sun, F., Zhang, G., Pan, X., & Zhou, Q. (2023). Long noncoding RNA PCGEM1 facilitates tumor growth and metastasis of osteosarcoma by sponging miR-433-3p and targeting OMA1. *Orthopaedic Audio-synopsis Continuing Medical Education, 15*, 1060–1071. https://doi.org/10.1111/os.13648.

Liao, W., Chu, P., Su, C., Wu, C., & Li, C. (2022). Mitochondrial AAA protease gene associated with immune infiltration is a prognostic biomarker in human ovarian cancer. *Pathology – Research and Practice, 240*, 154215. https://doi.org/10.1016/j.prp.2022.154215.

Luchsinger, L. L., de Almeida, M. J., Corrigan, D. J., Mumau, M., & Snoeck, H.-W. (2016). Mitofusin 2 maintains haematopoietic stem cells with extensive lymphoid potential. *Nature, 529*, 528–531. https://doi.org/10.1038/nature16500.

López-Pelegrín, M., Cerdà-Costa, N., Martínez-Jiménez, F., Cintas-Pedrola, A., Canals, A., Peinado, J. R., ... Gomis-Rüth, F. X. (2013). A novel family of soluble minimal scaffolds provides structural insight into the catalytic domains of integral membrane metallopeptidases. *The Journal of Biological Chemistry, 288*, 21279–21294. https://doi.org/10.1074/jbc.M113.476580.

Ma, Y., Wang, L., & Jia, R. (2020). The role of mitochondrial dynamics in human cancers. *American Journal of Cancer Research, 10*, 1278–1293.

Mac Vicar, T., Ohba, Y., Nolte, H., Mayer, F. C., Tatsuta, T., Sprenger, H., ... Langer, T. (2019). Lipid signalling drives proteolytic rewiring of mitochondria by YME1L. *Nature, 575*, 361–365. https://doi.org/10.1038/s41586-019-1738-6.

MacVicar, T. D. B., & Lane, J. D. (2014). Impaired OMA1-dependent cleavage of OPA1 and reduced DRP1 fission activity combine to prevent mitophagy in cells that are dependent on oxidative phosphorylation. *Journal of Cell Science, 127*, 2313–2325. https://doi.org/10.1242/jcs.144337.

Maycotte, P., Marín-hernández, A., Goyri-aguirre, M., Anaya-ruiz, M., Reyes-leyva, J., & Cortés-hernández, P. (2017). Mitochondrial dynamics and cancer. *Tumor Biology*, 1–16. https://doi.org/10.1177/1010428317698391.

Miallot, R., Millet, V., Groult, Y., Modelska, A., Crescence, L., Roulland, S., ... Naquet, P. (2023). An OMA1 redox site controls mitochondrial homeostasis, sarcoma growth, and immunogenicity. *Life Science Alliance, 6*, 1–17. https://doi.org/10.26508/lsa.202201767.

Mitra, K., Wunder, C., Roysam, B., Lin, G., & Lippincott-Schwartz, J. (2009). A hyperfused mitochondrial state achieved at G1-S regulates cyclin E buildup and entry into S phase. *Proceedings of the National Academy of Sciences of the United States of America, 106*, 11960–11965. https://doi.org/10.1073/pnas.0904875106.

Moreno-sánchez, R., Marín-hernández, A., Saavedra, E., Pardo, J. P., Ralph, S. J., & Rodríguez-enríquez, S. (2014). Who controls the ATP supply in cancer cells? Biochemistry lessons to understand cancer energy metabolism. *The International Journal of Biochemistry & Cell Biology, 50*, 10–23. https://doi.org/10.1016/j.biocel.2014.01.025.

Nassar, D., & Blanpain, C. (2016). Cancer stem cells: Basic concepts and therapeutic implications. *Annual Review of Pathology, 11*, 47–76. https://doi.org/10.1146/annurev-pathol-012615-044438.

Patten, D. A., Wong, J., Khacho, M., Soubannier, V., Mailloux, R. J., Pilon-Larose, K., ... Slack, R. S. (2014). OPA1-dependent cristae modulation is essential for cellular adaptation to metabolic demand. *The EMBO Journal, 33*, 2676–2691. https://doi.org/10.15252/embj.201488349.

Ponsoda, X., Bort, R., Ferrer-lorente, R., Prieto, J., Leo, M., & Torres, J. (2016). Early ERK1/2 activation promotes DRP1-dependent mitochondrial fission necessary for cell reprogramming. *Nature Communications, 7*(1), 13. https://doi.org/10.1038/ncomms11124.

Prieto, J., León, M., Ponsoda, X., Sendra, R., Bort, R., Ferrer-Lorente, R., ... Torres, J. (2016). Early ERK1/2 activation promotes DRP1-dependent mitochondrial fission necessary for cell reprogramming. *Nature Communications, 7*, 11124. https://doi.org/10.1038/ncomms11124.

Pyakurel, A., Savoia, C., Scorrano, L., Pyakurel, A., Savoia, C., Hess, D., & Scorrano, L. (2015). Extracellular regulated kinase phosphorylates mitofusin 1 to control mitochondrial morphology and apoptosis. *Molecular Cell, 58*, 244–254. https://doi.org/10.1016/j.molcel.2015.02.021.

Qian, W., Wang, J., Roginskaya, V., McDermott, L. A., Edwards, R. P., Stolz, D. B., ... Houten, B. Van (2014). Novel combination of mitochondrial division inhibitor 1 (mdivi-1) and platinum agents produces synergistic pro-apoptotic effect in drug resistant tumor cells. *Oncotarget, 5*, 4180–4194. https://doi.org/10.18632/oncotarget.1944.

Quirós, P. M., Ramsay, A. J., Sala, D., Fernández-Vizarra, E., Rodríguez, F., Peinado, J. R., ... López-Otín, C. (2012). Loss of mitochondrial protease OMA1 alters processing of the GTPase OPA1 and causes obesity and defective thermogenesis in mice. *The EMBO Journal, 31*, 2117–2133. https://doi.org/10.1038/emboj.2012.70.

Rainbolt, T. K., Lebeau, J., Puchades, C., & Wiseman, R. L. (2016). Reciprocal degradation of YME1L and OMA1 adapts mitochondrial proteolytic activity during stress. *Cell Reports, 14*, 2014–2049. https://doi.org/10.1016/j.celrep.2016.02.011.

Rivera-Mejías, P., Narbona-Pérez, Á. J., Hasberg, L., Kroczek, L., Bahat, A., Lawo, S., ... Langer, T. (2023). The mitochondrial protease OMA1 acts as a metabolic safeguard upon nuclear DNA damage. *Cell Reports, 42*. https://doi.org/10.1016/j.celrep.2023.112332.

Rodrigues, T., & Ferraz, L. S. (2020). Therapeutic potential of targeting mitochondrial dynamics in cancer. *Biochemical Pharmacology, 182*, 114282. https://doi.org/10.1016/j.bcp.2020.114282.

Roy, M., Reddy, P. H., Iijima, M., & Sesaki, H. (2015). Mitochondrial division and fusion in metabolism. *Current Opinion in Cell Biology, 33*, 111–118. https://doi.org/10.1016/j.ceb.2015.02.001.

Schwarzer, A., Oliveira, M., & Kleppa, M. (2023). Targeting AggressiveB-cell lymphomas through pharmacological activation of the mitochondrial protease OMA1. *Molecular Cancer Therapeutics, 22*(11), 1290–1303. https://doi.org/10.1158/1535-7163.MCT-22-0718.

Serasinghe, M. N., Wieder, S. Y., Renault, T. T., Elkholi, R., Asciolla, J. J., Yao, J. L., ... Chipuk, J. E. (2015). Mitochondrial division is requisite to RAS-induced transformation and targeted by oncogenic MAPK pathway inhibitors. *Molecular Cell, 57*, 521–536. https://doi.org/10.1016/j.molcel.2015.01.003.

Silic-Benussi, M., Scattolin, G., Cavallari, I., Minuzzo, S., del Bianco, P., Francescato, S., ... Ciminale, V. (2018). Selective killing of human T-ALL cells: an integrated approach targeting redox homeostasis and the OMA1/OPA1 axis. *Cell Death & Disease, 9*. https://doi.org/10.1038/s41419-018-0870-9.

Song, Z., Chen, H., Fiket, M., Alexander, C., & Chan, D. C. (2007). OPA1 processing controls mitochondrial fusion and is regulated by mRNA splicing, membrane potential, and Yme1L. *JCB, 178*, 749–755. https://doi.org/10.1083/jcb.200704110.

Srinivasainagendra, V., Sandel, M. W., Singh, B., Sundaresan, A., Mooga, V. P., Bajpai, P., ... Singh, K. K. (2017). Migration of mitochondrial DNA in the nuclear genome of colorectal adenocarcinoma. *Genome Medicine, 9*(1), 15. https://doi.org/10.1186/s13073-017-0420-6.

Suen, D., Norris, K. L., & Youle, R. J. (2008). Mitochondrial dynamics and apoptosis. *Genes & Development, 22*, 1577–1590. https://doi.org/10.1101/gad.1658508.

Uhlén, M., Fagerberg, L., Hallström, B. M., Lindskog, C., Oksvold, P., Mardinoglu, A., ... Pontén, F. (2015). Tissue-based map of the human proteome. *Science (80-), 347*, 1–11. https://doi.org/10.1126/science.1260419.

Varanita, T., Soriano, M. E., Sandri, M., Scorrano, L., Costa, V., Civiletto, G., ... Scorrano, L. (2015). The Opa1-dependent mitochondrial cristae remodeling pathway controls atrophic, apoptotic, and ischemic tissue damage. *Cell Metabolism, 21*, 834–844. https://doi.org/10.1016/j.cmet.2015.05.007.

Viana, M. P., Levytskyy, R. M., Anand, R., Reichert, A. S., & Khalimonchuk, O. (2021). Protease OMA1 modulates mitochondrial bioenergetics and ultrastructure through dynamic association with MICOS complex. *IScience, 24*, 102119. https://doi.org/10.1016/j.isci.2021.102119.

Wai, T., & Langer, T. (2016). Mitochondrial dynamics and metabolic regulation. *Trends in Endocrinology and Metabolism, 27*, 105–117. https://doi.org/10.1016/j.tem.2015.12.001.

Wai, T., García-Prieto, J., Baker, M. J., Merkwirth, C., Benit, P., Rustin, P., ... Langer, T. (2015). Imbalanced OPA1 processing and mitochondrial fragmentation cause heart failure in mice. *Science (80-), 350*. https://doi.org/10.1126/science.aad0116.

Warburg, O. (1956). On the origin of cancer cells. *Science (80-), 123*, 309–314.

Westermann, B. (2012). Bioenergetic role of mitochondrial fusion and fission. *Biochimica et Biophysica Acta, 1817*, 1833–1838. https://doi.org/10.1016/j.bbabio.2012.02.033.

Winter, J. M., Yadav, T., & Rutter, J. (2022). Stressed to death: Mitochondrial stress responses connect respiration and apoptosis in cancer. *Molecular Cell, 82*, 1–12. https://doi.org/10.1016/j.molcel.2022.07.012.

Wu, Z., Zuo, M., Zeng, L., Cui, K., Liu, B., Yan, C., ... Song, Z. (2021). OMA1 reprograms metabolism under hypoxia to promote colorectal cancer development. *EMBO Reports, 22*, 1–18. https://doi.org/10.15252/embr.202050827.

Xie, Q., Wu, Q., Horbinski, C. M., Flavahan, W. A., Yang, K., Zhou, W., ... Rich, J. N. (2015). Mitochondrial control by DRP1 in brain tumor initiating cells. *Nature Neuroscience, 18*, 501–510. https://doi.org/10.1038/nn.3960.

Xie, B., Wang, S., Jiang, N., & Li, J. J. (2019). Cyclin B1/CDK1-regulated mitochondrial bioenergetics in cell cycle progression and tumor resistance. *Cancer Letters, 443*, 56–66. https://doi.org/10.1016/j.canlet.2018.11.019.

Xing, J., Qi, L., Liu, X., Shi, G., Sun, X., & Yang, Y. (2022). Roles of mitochondrial fusion and fission in breast cancer progression: A systematic review. *World Journal of Surgical Oncology, 5*, 1–13.

Yang, F., Wu, R., Jiang, Z., Chen, J., Nan, J., Su, S., ... Wang, J. (2018). Leptin increases mitochondrial OPA1 via GSK3-mediated OMA1 ubiquitination to enhance therapeutic effects of mesenchymal stem cell transplantation. *Cell Death & Disease, 9*. https://doi.org/10.1038/s41419-018-0579-9.

Yuen, K. W. Y., Warren, C. D., Chen, O., Kwok, T., Hieter, P., & Spencer, F. A. (2007). Systematic genome instability screens in yeast and their potential relevance to cancer. *Proceedings of the National Academy of Sciences of the United States of America, 104*, 3925–3930. https://doi.org/10.1073/pnas.0610642104.

Zhang, G., Jin, H., Lin, X., Chen, C., Liu, X., & Zhang, Q. (2013). Anti-tumor effects of Mfn2 in gastric cancer. *International Journal of Molecular Sciences,* 13005–13021. https://doi.org/10.3390/ijms140713005.

Zhang, K., Li, H., & Song, Z. (2014). Membrane depolarization activates the mitochondrial protease OMA1 by stimulating self-cleavage. *EMBO Reports, 15*, 576–585. https://doi.org/10.1002/embr.201338240.

Zhao, J., Zhang, J., Yu, M., Xie, Y., Huang, Y., Wolff, D. W., ... Tu, Y. (2013). Mitochondrial dynamics regulates migration and invasion of breast cancer cells. *Oncogene, 32*, 4814–4824. https://doi.org/10.1038/onc.2012.494.

Zunino, R., Braschi, E., Xu, L., & McBride, H. M. (2009). Translocation of SenP5 from the nucleoli to the mitochondria modulates DRP1-dependent fission during mitosis. *The Journal of Biological Chemistry, 284*, 17783–17795. https://doi.org/10.1074/jbc.M901902200.

CHAPTER FOUR

Role of antioxidants in modulating anti-tumor T cell immune resposne

Nathaniel Oberholtzer, Stephanie Mills, Shubham Mehta, Paramita Chakraborty, and Shikhar Mehrotra*

Department of Surgery, Hollings Cancer Center, Medical University of South Carolina, Charleston, SC, United States
*Corresponding author. e-mail address: mehrotr@musc.edu

Contents

1. Introduction	100
1.1 ROS production by T cells	100
1.2 Role of ROS in T cell signaling	101
1.3 Negative impact of ROS on T cell function	102
1.4 Antioxidant defense mechanisms in T cells	105
1.5 Sources of ROS in the TME	109
1.6 Antioxidant molecules and metabolites targeted in anti-tumor immunity	115
1.7 Clinical trials targeting immune cell antioxidant metabolism	115
1.8 Conclusions and future directions	115
Acknowledgment	116
References	116

Abstract

It has been well established that in addition to oxygen's vital in cellular respiration, a disruption of oxygen balance can lead to increased stress and oxidative injury. Similarly, reduced oxygen during tumor proliferation and invasion generates a hypoxic tumor microenvironment, resulting in dysfunction of immune cells and providing a conducive milieu for tumors to adapt and grow. Strategies to improve the persistence tumor reactive T cells in the highly oxidative tumor environment are being pursued for enhancing immunotherapy outcomes. To this end, we have focused on various strategies that can help increase or maintain the antioxidant capacity of T cells, thus reducing their susceptibility to oxidative stress/damage. Herein we lay out an overview on the role of oxygen in T cell signaling and how pathways regulating oxidative stress or antioxidant signaling can be targeted to enhance immunotherapeutic approaches for cancer treatment.

Advances in Cancer Research, Volume 162
ISSN 0065-230X, https://doi.org/10.1016/bs.acr.2024.05.003
Copyright © 2024 Elsevier Inc. All rights are reserved, including those for text and data mining, AI training, and similar technologies.

1. Introduction

Since the original experiments that identified that specific antigens associated with tumors could be recognized by immune cells, significant advances have been made in the field of cancer immunotherapy to improve efficacy, including the use of high-dose IL-2 therapy, cloning tumor antigens, identifying tumor-reactive T cells, and engineering T cells to have direct anti-tumor specificity (Cheever, Greenberg, & Irle, 1984; Greenberg, Cheever, & Fefer, 1981; Kawakami & Rosenberg, 1996; Old, 1982; Topalian, Hom, & Kawakami, 1992). The initial cloning of the T cell receptor (TCR) and demonstration of its successful engineering in T cells led to the revolutionizing approach of transferring patient T cells that had been genetically modified to recognize tumor antigens, termed adoptive T cell therapy (ACT) (Clay et al., 1999; Duval, Schmidt, & Kaltoft, 2006; Morgan, Dudley, & Wunderlich, 2006). These strategies have been enhanced by the use of chimeric antigen receptors (CARs) to direct T cell specificity and by blocking immune checkpoint molecules to reduce T cell anergy and immune evasion.

While these strategies have shown significant promise and have led to complete and durable regression in some patients, the obstacles posed by the highly immunosuppressive tumor microenvironment (TME) have led to limited overall efficacy, especially in solid tumors (Anderson, Stromnes, & Greenberg, 2017). One unique characteristic of the TME is elevation in reactive oxygen species (ROS), which have both positive and negative effects on T cells (Weinberg, Ramnath, & Nagrath, 2019). At low, physiological conditions, ROS have been shown to be critical for T cell activation (Sena, Li, & Jairaman, 2013a). However, at the sustained, high levels of ROS observed in the TME, infiltrating T cells experience oxidative stress, which ultimately leads to dysfunction and lack of persistence (Otsuji, Kimura, Aoe, Okamoto, & Saito, 1996). Thus, strategies to boost the antioxidant capacity of T cells to reduce their susceptibility to oxidative stress in the TME have shown promise in preclinical models. Here, we discuss the role that ROS play in T cell activation and function, the negative impacts of ROS on T cell function, the adaptations that tumor cells have developed to produce high levels of ROS, and the strategies that have been explored to target oxidative stress in cancer immunotherapy.

1.1 ROS production by T cells

In recent years, there has been a growing emphasis on studying the role of antioxidant metabolism in regulating the phenotype and function of immune cells, particularly T lymphocytes. Importantly, the high metabolic demand of activated T cells leads to upregulation of several pathways to support their effector functions. These metabolic pathways tend to drive production of reactive oxygen species (ROS), which serve as both critical signaling molecules as well as cytotoxic agents that, in excess, inhibit the effector function of T cells (Flescher, Ledbetter, & Schieven, 1994; Flescher, Tripoli, Salnikow, & Burns, 1998; Kamiński, Sauer, & Kamiński, 2012a; Sena, Li, & Jairaman, 2013b). Two major metabolic pathways drive ROS production in T cells: oxidative phosphorylation (OXPHOS, the primary source of ROS) and the tricarboxylic acid (TCA) cycle. In OXPHOS, electrons are shuttled along the electron transport chain (ETC) to ultimately drive ATP synthesis. However, during times of high metabolic demand, electrons can leak through the complexes of the ETC where they can react with and reduce O_2 to generate superoxide radicals (O_2^-) and hydrogen peroxide (H_2O_2) (Zhao, Jiang, Zhang, & Yu, 2019). In addition to ETC-driven ROS production, the TCA cycle has also been implicated in ROS generation through the activity of three main enzymes: pyruvate dehydrogenase (PDH) alpha-ketoglutarate dehydrogenase (a-KGDH) and succinate dehydrogenase (SDH). PDH and a-KGDH generate hydrogen peroxide or superoxide radicals upon perturbation of nicotinamide adenine dinucleotide (NAD+) levels while succinate dehydrogenase mediates radical production by reverse electron transport through Complex I (Chouchani, Pell, & Gaude, 2014; O'Brien, Chalker, Slade, Gardiner, & Mailloux, 2017; Tretter & Adam-Vizi, 2004; Zhang, Zhang, & Zhu, 2020). Beyond these primary metabolic sources, NADPH oxidase (NOX) enzymes are key sources of cytoplasmic ROS in T cells (Chen, Liu, & Chernatynskaya, 2024; Tse, Thayer, & Steele, 2010). Finally, TCR stimulation has also been shown to induce both NOX- and OXPHOS-derived ROS generation (Jackson, Devadas, Kwon, Pinto, & Williams, 2004; Kamiński et al., 2012a).

1.2 Role of ROS in T cell signaling

The acute induction of ROS production upon TCR stimulation suggests a potential role of ROS as a "second signal" capable of influencing fate and function of activated T cells. Indeed, multiple studies have demonstrated the necessity of ROS signaling in supporting T cell activation (Kamiński et al., 2012a; Sena et al., 2013b). One mechanism by which acute ROS signaling supports T cell activation is through activation of ROS-dependent transcription factors, including NF-kB and AP-1 (Kaminski, Sauer, & Klemke, 2010; Kamiński et al., 2012b). Sena et al. further demonstrated that mitochondria-derived ROS production is required for activation of nuclear actor of activated T cells (NFAT) and subsequent IL-2 production, and that T cells lacking mitochondrial ROS production were unable to undergo antigen-induced proliferation both in vitro and in vivo (Sena et al., 2013b). A well-established feature of effector T cells is that they tend to undergo a metabolic switch from OXPHOS to aerobic glycolysis upon T cells stimulation (Frauwirth, Riley, & Harris, 2002). Induction of glycolytic enzymes following TCR stimulation has been shown to mediate mitochondrial ROS production, which in turn increases IL-2 production (Kamiński et al., 2012a). These studies have demonstrated that TCR stimulation-induced ROS is indispensable for T cell activation, and that ROS production and cellular metabolism are intimately linked in T cells.

Studies have also demonstrated that T cell subsets are differentially regulated by ROS signaling, with opposing effects observed in Th1 versus Th2 cells (Frossi, De Carli, Piemonte, & Pucillo, 2008; Gerriets, Kishton, & Nichols, 2015). Specifically, treatment of Th1 and Th2 cells with low doses of H_2O_2 reduced production of IFNg in the Th1 subset and increased IL-4 production in the Th2 subset (Frossi et al., 2008). A similar effect of ROS on IL-4 production was observed by Kaminski et al., who demonstrated that mitochondria-derived ROS is critical for activation-induced IL-2 and IL-4 production upon TCR stimulation (Kaminski et al., 2010). These findings suggest that ROS signaling preferentially enhances the Th2 response, which could have important implications for allergic diseases as well as the anti-tumor immune response. Other studies have shown a role for ROS in modulating the Th17 response (Tse et al., 2010). When superoxide production was inhibited by knocking out NADPH oxidase (NOX), activation skewed T cells toward a Th17 phenotype, whereas NOX-dependent superoxide production resulted in a Th1 phenotype (Tse et al., 2010).

1.3 Negative impact of ROS on T cell function

While the findings described above indicate a necessary role of ROS in supporting T cell activation and function, it is also understood that excessively high and sustained levels of ROS can ultimately be detrimental to T cell survival. When ROS levels are sufficiently high to overwhelm antioxidant regulatory systems, ROS-induced damage of nucleic acids, proteins, and lipids occurs, resulting in oxidative stress (Ray, Huang, & Tsuji, 2012). ROS react with DNA to induce double stranded breaks, leading to genomic instability (Jeggo & Löbrich, 2007). When reacting with proteins, ROS can oxidize thiol-containing cysteine and methionine residues, leading to protein misfolding and dysfunction (Nakamura, Oh, Zhang, & Lipton, 2021; Shields, Traa, & Van Raamsdonk, 2021). ROS can also induce lipid peroxidation, which results in damage to lipid membranes and produces lipid peroxide radicals (Shields et al., 2021; Tribble, Aw, & Jones, 1987). In T cells, tumor-induced oxidative stress has been shown to increase susceptibility to TNFα-mediated apoptosis through inhibition of NF-κβ signaling (Bhattacharyya, Mandal, & Sen, 2007). Additionally, neutrophil-derived ROS have been shown to inhibit DNA synthesis and limit effector function in T cells (Cemerski, Cantagrel, van Meerwijk, & Romagnoli, 2002). This decrease in effector function was linked to ROS-dependent perturbations in the TCR signaling pathway (Cemerski et al., 2002; Mougiakakos, Johansson, & Kiessling, 2009). Other studies have demonstrated that sustained exposure of T cells to oxidative stress leads to suppression of calcium flux, NFAT and NFkβ signaling, and IL-2 production (Flescher et al., 1994). Interestingly, the enzyme glutathione peroxidase 4 (Gpx4), one of the main enzymes responsible for scavenger phospholipid hydroperoxides and combating lipid peroxidation, is critically important for maintenance of T cell immune responses in infection, suggesting that oxidative stress-induced lipid peroxidation impairs normal T cell function (Matsushita et al., 2015). Similar observations have been made in the setting of the TME where uptake of oxidized lipids by CD8 + TILs induces lipid peroxidation and T cell dysfunction, which could be reversed by overexpression of Gpx4 or blockage of type I interferon signaling (Chen, Teo, & Yau, 2022; Xu, Chaudhary, & Rodríguez-Morales, 2021a).

Direct treatment of T cells with H_2O_2 induces apoptosis to varying degrees depending on the T cell subset. Low doses of H_2O_2 preferentially induced apoptosis in effector memory T cells (T_{em}) via a mechanism dependent on mitochondrial depolarization and caspase 3 activation (Takahashi, Hanson, & Norell, 2005). Conversely, low doses of H_2O_2 appear to have no impact on viability or suppressive capacity of regulatory T cells (T_{reg}) (Mougiakakos et al., 2009). Similarly, naive T cells appear to have minimal sensitivity to oxidative stress (Takahashi et al., 2005). Thus, oxidative stress does not impact all T cells equally, an important concept to further explore when developing redox-targeted therapies in the context of immunotherapy for cancer.

ROS have also been implicated in mediating activation-induced cell death (AICD), the process by which antigen-experienced T cells undergo apoptosis upon TCR stimulation. AICD has been identified as a contributing factor to immune-escape in the setting of anti-tumor immunity, as multiple studies have demonstrated that tumor cells induce Fas-FasL interactions with T cells to promote AICD (Cao, Wang, & Li, 2015; Hahne, Rimoldi, & Schröter, 1996). While not fully understood, it is thought that ROS produced through TCR activation leads to increased cytoplasmic Ca^{2+}, which activates calcineurin and ultimately leads to downstream nuclear translocation of NFAT (Hildeman, Mitchell, Kappler, & Marrack, 2003; Schieven, Mittler, & Nadler, 1994). NFAT then drives expression of FasL to increase T cell sensitivity to AICD (Rengarajan, Mittelstadt, & Mages, 2000; Shin, Kim, Cho, & Nguyen, 2019). The role of ROS in regulating FasL and thus AICD is supported by studies showing that treatment of T cells with H_2O_2 in the absence of TCR stimulation induces FasL expression (Bauer et al., 1998; Devadas, Zaritskaya, Rhee, Oberley, & Williams, 2002). In addition to AICD, activated T cell–autonomous death (ACAD), which is independent of FasL signaling, also appears to be regulated by ROS. One target of interest in ACAD, is antioxidant molecule Bcl-2 which was significantly upregulated by treatment with exogenous antioxidants, suggesting that ROS leads to cAMP response element-binding protein (CREB) mediated downregulation of Bcl-2 (Hildeman et al., 2003; Wilson, Mochon, & Boxer, 1996) Thus, evidence suggests that elevated ROS limits T cell effector functions in the TME through promotion of death pathways AICD and ACAD.

Similar to the difference in susceptibility to oxidative stress between T cell subsets, there is also a difference in susceptibility to AICD between T cell subsets (Kesarwani, Al-Khami, & Scurti, 2014; Mehrotra et al., 2004). Namely, our lab has shown that antigen-specific cytotoxic lymphocytes with a T_{em} phenotype are more susceptible to AICD compared to cytotoxic lymphocytes with a central memory (T_{cm}) phenotype (Mehrotra et al., 2004). These findings are of particular interest given the generally accepted superiority of T_{cm} cells in the context of anti-tumor immune control, with T_{em} cells being limited by poor persistence and reduced anti-tumor efficacy upon adoptive transfer (Kesarwani et al., 2014). Our lab established that T_{cm} cells are characterized by an increased in levels of cell surface thiols (c-SH) compared to T_{em}, which corresponded to an increase in overall antioxidant capacity and intracellular glutathione (iGSH) in the T_{cm} subset compared to T_{em} cells (Kesarwani et al., 2014). After proliferation on repeated TCR stimulation, T cells progressively lose c-SH expression (Kesarwani et al., 2014). Increasing c-SH expression on T cells using thiol donors like N-acetylcysteine (L-NAC) rescues T cells from TCR stimulation-induced AICD and confers superior anti-tumor capacity (Chakraborty, Chatterjee, & Kesarwani, 2019; Kesarwani et al., 2014). Collectively, these findings demonstrate variance in susceptibility to ROS-mediated AICD between T cell subsets with T_{em} cells being the most susceptible, and that strategies to enhance antioxidant capacity can protect T cells from AICD.

1.4 Antioxidant defense mechanisms in T cells

To develop strategies to protect anti-tumor T cells from oxidative stress in the TME, it is first important to understand the various mechanisms that regulate antioxidant capacity in T cells. Through epigenetic regulation of antioxidant genes, to regulation of antioxidant states of proteins, changes in antioxidant levels can protect T cells from dysfunction, alter their state of activation and promote anti-tumor killing capacities.

Nrf2: One of the main transcription factors responsible for regulating the expression of antioxidant genes is nuclear factor erythroid-2-related factor 2 (Nrf2), which is widely considered the "master regulator" of antioxidant responses (Vomund, Schäfer, Parnham, Brüne, & von Knethen, 2017). At baseline, Nrf2 levels are nearly undetectable in T cells; however, upon TCR stimulation, T cells dramatically upregulate expression of Nrf2, resulting in increased expression of Nrf2 target antioxidant enzymes (Morzadec et al., 2014). Several studies have demonstrated the importance of Nrf2 expression in maintaining functionality of T cells. One study found that the Th1 response declines in aging and that this decline could be prevented by treatment with an Nrf2 agonist or treatment with L-NAC (Kim, Barajas, Wang, & Nel, 2008). Other studies have shown that Nrf2 regulates T cell sensitivity to Fas-mediated apoptosis by regulating intracellular glutathione levels (Morito, Yoh, & Itoh, 2003). In this study, Nrf2-deficient T cells were more susceptible to Fas-mediated apoptosis and restoring levels of intracellular glutathione reversed this sensitivity to apoptosis (Morito et al., 2003). Collectively, these studies demonstrate the critical role Nrf2 plays in regulating antioxidant capacity in T cells to support their function and persistence. The individual antioxidant molecules and enzymes that protect T cells from oxidative stress, particularly those that have been implicated in anti-tumor immunity, will be explored next.

Glutathione (GSH) and thiols: GSH is a non-enzymatic tripeptide antioxidant abundant in mammalian cells which exists in two states: reduced (GSH) and oxidized (GSSG) (Pizzorno, 2014). Under normal physiological conditions, the reduced form is much more abundant than the oxidized form (up to 100-fold higher). Glutathione undergoes conversion between the reduced and oxidized form when GSH reacts with electrophilic compounds such as free radicals (Aquilano, Baldelli, & Ciriolo, 2014). Once oxidized, GSSG can be enzymatically reduced back to GSH by glutathione reductase (using NADPH as an electron donor) (Aquilano et al., 2014). In this way, the ratio of reduced to oxidized forms of glutathione serves as an important indicator of the overall redox status of a cell. GSH serves as an important antioxidant due to its ability to act as a scavenger of free radicals and ROS. The antioxidant function of GSH is largely dependent on its thiol moiety, which directly scavenges free radicals (Johnson, Wilson-Delfosse, & Mieyal, 2012). In addition to directly scavenging free radicals, GSH serves an important role in protecting protein thiols from oxidation. Protein thiols are critical modifications important for protein function given the highly conserved nature of cysteine residues (Ulrich & Jakob, 2019). Oxidation of these thiol groups can have negative impacts on protein function, thus tight regulation of these protein thiols by GSH and other thiol-containing compounds is critical to prevent negative impacts on protein function (Requejo, Hurd, Costa, & Murphy, 2010).

Our group has shown that global levels of cell surface thiols (c-SH) vary between T cells subsets, with the T_{em} subset expressing lower levels of c-SH (correlating to increased sensitivity to ACID and oxidative stress) and the T_{cm} subset expressing higher levels (correlating to higher antioxidant capacity and decreased sensitivity to AICD) (Kesarwani et al., 2014). Treatment of T cells with thiol donors (such as L-NAC or rapamycin) promote c-SH expression and enhanced anti-tumor capacity (Kesarwani et al., 2014). L-NAC treatment has also been shown to restore metabolic T cell function during chronic stimulation (Vardhana, Hwee, & Berisa, 2020). Further, our group found that when anti-tumor T cells were sorted based on high (c-SH^{hi}) and low (c-SH^{lo}) expression of surface thiols, the c-SH^{hi} subset was characterized by increased levels of iGSH, superior antioxidant capacity, and greater anti-tumor capacity when adoptively transferred into tumor bearing mice (Kesarwani et al., 2014). c-SH^{hi} cells also persisted longer in vivo after adoptive transfer, supporting our hypothesis that protein thiols play an important role in protecting T cells from oxidative stress and AICD (Kesarwani et al., 2014). Interestingly, we also observed an inverse relationship between L-NAC induced c-SH expression and mTOR activity. Treatment of T cells with rapamycin blocked mTOR activity and simultaneously increased c-SH expression and antioxidant capacity (Kesarwani et al., 2014). This is consistent with other observations that increased mTOR activity is associated with higher levels of ROS in T cells (Kesarwani et al., 2014). Of note, rapamycin treatment has been shown to induce expansion of Tregs in human patients, which correlates to the high level of reduced thiols present and robust antioxidant capacity in Tregs (Battaglia et al., 2006; Mougiakakos et al., 2009).

Thioredoxin (Trx): Thioredoxins are a family of redox proteins that are ubiquitously expressed and serve as one of the key antioxidant systems in cells (Lee, Kim, & Lee, 2013). Trx proteins have cysteine-containing redox-active centers that scavenge reactive species by catalyzing electron transfer from NADPH via thioredoxin reductase to Trx (Lee et al., 2013). Trx has been shown to protect T cells from oxidative stress induced apoptosis by its ability to scavenge H_2O_2 (Iwata, Hori, & Sato, 1997). Our group found that treatment of T cells with IL-15 (a cytokine known to promote the T_{cm} phenotype) induced expression of thioredoxin-1 (Trx1) and promoted T cell survival (Kesarwani et al., 2014). Similar to the observed increase in protein thiols in Tregs, human Tregs also expressed higher levels of Trx1 which conferred enhanced antioxidant capacity and reduced susceptibility to oxidative stress (Mougiakakos, Johansson, Jitschin, Böttcher, & Kiessling, 2011). Given these findings, our group tested the effect of overexpressing Trx1 in T cells on their anti-tumor capacity (Chakraborty et al., 2019). In these experiments, Pmel transgenic mice were designed to overexpress Trx1 (Pmel-Trx1). Pmel-Trx1 T cells had lower levels of ROS and decreased susceptibility to AICD. Pmel-Trx1 cells were also characterized by a T_{cm} phenotype with a distinct metabolic profile and enhanced mitochondrial spare respiratory capacity (Chakraborty et al., 2019). Further, treatment of Pmel T cells with recombinant Trx1 prior to adoptive transfer into melanoma tumor-bearing mice resulted in superior persistence and tumor control (Chakraborty et al., 2019).

Catalase: Catalase is an important scavenger of ROS in cells through its ability to enzymatically convert H_2O_2 to H_2O and O_2. The importance of catalase in preventing apoptosis in T cells is well established (Sandstrom & Buttke, 1993). Studies have shown that overexpressing catalase in T cells via retroviral transduction protects T cells from ROS-mediated damage and enhances the survival of T cells during oxidative stress (Ando, Mimura, & Johansson, 2008). This strategy has produced similar results in CAR-T cells, with T cells engineered to express catalase along with the CAR showing increased levels of intracellular catalase, superior antioxidant capacity, and enhance anti-tumor capacity upon adoptive transfer (Ligtenberg, Mougiakakos, & Mukhopadhyay, 2016). Interestingly, CAR-T co-expressing catalase also produced a protective bystander effect, whereby non-transfected immune cells were also protected from oxidative stress when mixed with catalase expressing CAR-T cells (Ligtenberg et al., 2016). Additionally, recent studies have further confirmed the protective effect of catalase in T cells by showing that simply treating T cells with recombinant catalase reduces T cell exhaustion, promotes the T_{cm} phenotype, and supports T cell persistence and effector function upon adoptive transfer (Aksoylar & Patsoukis, 2023).

Glutathione peroxidase (GPX): Similar to catalase, GPX enzymes are involved in reducing excess H_2O_2 to protect cells from oxidative damage. This protect effect has been shown to be critical for T cell function, as GPX1 plays a role in reducing ROS production following TCR stimulation and protecting T cells from ROS-induced apoptosis (Lee et al., 2016; Won, Sohn, & Min, 2010). In Tregs, GPX4 has been shown to play an important role in limiting peroxidation and ferroptosis to sustain Treg function and suppressive capacity in the tumor microenvironment (Xu, Sun, & Johnson, 2021b). Further studies are needed to fully elucidate the role of GPX enzymes in regulating cytotoxic and regulatory T cells in the context of anti-tumor immunity.

Superoxide dismutase (SOD): Akin to catalase, SODs are a family of enzymes that catalyze the dismutation of two superoxide species, producing a molecule of O_2 and a molecule of H_2O_2, which is then metabolized by catalase along with other antioxidant enzymes. Three main isoforms of SOD exist: SOD1 (cytoplasmic), SOD2 (mitochondrial), and SOD3 (extracellular) (Fukai & Ushio-Fukai, 2011). SOD2 has been implicated as one of the primary defense mechanisms against free radical mediated damage in T cells (Pardo, Melendez, & Tirosh, 2006). Using a T cell-specific SOD2 conditional knockout model, lack of SOD2 was found to result in an increase in accumulation of superoxide species, developmental defects in T cells, and impaired T cell immune response (Case, McGill, & Tygrett, 2011). Studies have also shown that Fas-mediated apoptosis induces signaling that selectively degrades SOD2 and inactivates its antioxidant activity (Pardo et al., 2006). Thus, loss of SOD2 activity in T cells upon Fas ligation increases accumulation of mito-chondrial superoxide species, impairs mitochondrial function, and potentiates apoptosis (Pardo et al., 2006). Our group found that treatment of tumor antigen-specific T cells with the SOD mimetic MnTBAP protected cells from AICD while preserving effector function (Norell, Martins da Palma, & Lesher, 2009). Similar functional benefits were observed in TILs when treated *ex vivo* with MnTBAP (Norell et al., 2009). Collectively, these findings demonstrate the importance of SOD in maintaining redox balance in T cells and provide a potential therapeutic strategy for targeting SOD in anti-tumor T cells.

Nicotinamide adenine dinucleotide (NAD+): NAD+ (and its reduced form NADH) serves as an important co-factor in many metabolic and redox reactions (Morandi, Horenstein, & Malavasi, 2021). In fact, the intracellular ratio of NAD+ /NADH serves as a major indicator of the redox status of a cell, with declining NAD+ /NADH ratios correlating to a decrease in overall antioxidant capacity. One of the main targets of NAD+ in antioxidant metabolism is the NAD+-dependent sirtuin SIRT3 (Someya, Yu, & Hallows, 2010). SIRT3-mediated deacetylation of the mitochondrial enzyme Idh2 leads to increase in NADPH and a subsequent increase in the GSH/GSSG ratio (Someya et al., 2010). NAD+-dependent deacetylation by SIRT3 has also been shown to enhance activation of key antioxidant enzymes including catalase and SOD (Sundaresan et al., 2009). In the setting of anti-tumor immunity, our group has shown that NAD+-dependent activity of SIRT1 enhances anti-tumor potential of adoptively transferred T cells but is limited by the NADase activity of CD38 which depletes intracellular NAD+ levels (Chatterjee, Daenthanasanmak, & Chakraborty, 2018). Subsequent blocking CD38 expression in T cells leads to increased levels of NAD+ and restores anti-tumor function (Chatterjee et al., 2018). Subsequent studies have shown that the levels of the main enzyme responsible for producing NAD+ in the salvage pathway, nicotinamide phosphoribosyltransferase (NAMPT), is required for T cell activation and that NAMPT is expressed at lower levels in tumor infiltrating lymphocytes (Wang, Wang, & Wang, 2021). NAD+ depletion in TILs led to disruptions in T cell metabolism and anti-tumor effector function. Conversely, supplementation with NAD+ led to enhanced tumor killing capacity in T cells in models of both CAR-T and immune checkpoint therapy (Wang et al., 2021). Treatment with nicotinamide (NAM) and nicotinamide riboside (NR) has also been shown to increase levels of NAD+, reverse mitochondrial ROS accumulation, and mitigate exhaustion in antitumor T cells (Yu, Imrichova, & Wang, 2020). Collectively, these studies demonstrate the importance of NAD+ as a redox regulator in T cells which can be therapeutically modulated to enhance immunotherapeutic tumor control.

1.5 Sources of ROS in the TME

The TME is characterized by a host of factors that promote tumor cell growth and disrupt the anti-tumor immune response. One of these factors is elevated levels of ROS, which are produced by both tumor cells themselves and the pro-tumorigenic cell types that make up the TME (Martinez-Outschoorn, Balliet, & Rivadeneira, 2010; Szatrowski & Nathan, 1991;

Weinberg, Hamanaka, & Wheaton, 2010). In this section, we outline the cells that make up the TME and their contributions to ROS production.

Tumor cells: Elevated ROS production is characteristic of most human cancer cells (Szatrowski & Nathan, 1991). The primary source of ROS generation in tumor cells is from the mitochondria as oxygen is reduced through the electron transport chain (ETC) (Weinberg et al., 2010, 2019). Complexes I, II, and III are significant sites of ROS generation with the majority of ROS leaking from the mitochondria into the cytoplasmic space from complex III (Muller, Liu, & Van Remmen, 2004; Orr, Vargas, & Turk, 2015; Turrens, 2003; Weinberg et al., 2019). ROS are also known to interact with DNA which leads to DNA damage and generation of oncogenic mutations and genomic instability in cancer cells (Srinivas, Tan, Vellayappan, & Jeyasekharan, 2019). In addition to induction of genomic instability and generation of oncogenes, ROS production has been shown to play an important role in tumorigenesis and progression through stimulation of various tumorigenic signaling pathways. One of the most well-established roles of ROS in tumor cells is their role in stabilizing hypoxia-inducible transcription factors (HIFs), which are critical for cell survival under the hypoxic conditions of the TME (Chandel et al., 1998; Wang, Jiang, Rue, & Semenza, 1995; Weinberg et al., 2019). Specifically, it has been shown that ROS generated from complex III of the ETC both induce expression of HIF-1a and stabilizes its expression under conditions of hypoxia (Chandel et al., 1998). Stabilized HIF expression induces transcriptional programming that promotes tumor cell growth and angiogenesis, another key characteristic of tumorigenesis (Pugh & Ratcliffe, 2003). ROS have also been shown to influence components of the MAPK signaling pathways to promote tumor cell proliferation, including inhibition of MAPK phosphatases and activation of EGFR through the RAS and ERK pathways (Messina, De Simone, & Ascenzi, 2019; Meves, Stock, Beyerle, Pittelkow, & Peus, 2001; Park, Nam, & Yun, 2011; Son et al., 2011). The PI3K/AKT/mTOR pathway is another important signaling pathway involved in tumorigenesis and metastasis that is hyperactivated by ROS through oxidation of thiol groups on various phosphatases within the pathway (Kirtonia, Sethi, & Garg, 2020; Lee et al., 2002; Salmeen, Andersen, & Myers, 2003). ROS also stimulates the transcription factor NF-κβ in cancer cells (Li & Engelhardt, 2006; Morgan & Liu, 2011; Ruiz-Ramos, Lopez-Carrillo, Rios-Perez, De Vizcaya-Ruíz, & Cebrian, 2009). Of note, NF-κβ signaling has been demonstrated to be critical for tumorigenesis, including roles in regulating proliferation, migration, and apoptosis of cancer tumors (Dolcet, Llobet, Pallares, & Matias-Guiu, 2005). A hallmark of cancer cells is a metabolic shift to aerobic glycolysis, termed the Warburg effect, and studies have shown that SOD2 upregulation in cancer cells sustains this metabolic shift via mitochondrial ROS that stimulates AMPK signaling (Hart, Mao, & de Abreu, 2015). Collectively, these findings demonstrate the intricate and interconnected mechanisms that cancer cells have evolved to use ROS to their benefit to the detriment of anti-tumor immune cells.

Cancer associated fibroblasts (CAFs): CAFs are a major component of the TME and have been shown to play a role in cancer progression and metastasis (Orimo, Gupta, & Sgroi, 2005). In addition to being stimulated by elevated levels of ROS produced by tumor cells, CAFs have also been shown to produce ROS themselves (Balliet, Capparelli, & Guido, 2011; Hanley, Mellone, & Ford, 2018; Ippolito, Morandi, & Taddei, 2019). CAF differentiation is dependent on generation of ROS by NOX4, which correlates to an increase in NOX4 expression in multiple human cancers. This increase in NOX4 expression is also strongly correlated with increased frequencies of CAFs in human tumors (Hanley et al., 2018). Interestingly, inhibition of NOX4 has been shown to prevent CAF accumulation in tumors and slow tumor growth (Hanley et al., 2018). CAFs have also been shown to play a role in tumor progression by inducing ROS-generating myeloid-derived suppressor cells (MDSCs) which dampen the anti-tumor immune response (Xiang, Ramil, & Hai, 2020).

Myeloid-derived suppressor cells (MDSCs): MDSCs are a subset of immunosuppressive myeloid cells found at elevated levels in nearly all human tumors (Ostrand-Rosenberg, Sinha, Beury, & Clements, 2012). MDSCs both induce angiogenesis to promote tumor metastasis and inhibit anti-tumor immunity through a variety of mechanisms, including production of ROS and nitric oxide species (NOS) (Ezernitchi, Vaknin, & Cohen-Daniel, 2006; Mazzoni, Bronte, & Visintin, 2002; Rice, Davies, & Subleski, 2018; Yang, DeBusk, & Fukuda, 2004). In tumor bearing mice, MDSC suppression of T cells was dependent on NOX2 mediated ROS production controlled by the STAT3 transcription factor (Corzo, Cotter, & Cheng, 2009). It has been shown that ROS from MDSCs can directly disrupt the ability of the TCR-CD8 complex to bind to cancer peptide-MHC complexes, leading to an inability of anti-tumor T cells to recognize tumor antigens (Nagaraj, Gupta, & Pisarev, 2007). Indirectly, MDSCs can also suppress T cell activation by depleting extracellular levels of cysteine, an important precursor to antioxidant glutathione production (Srivastava, Sinha, Clements, Rodriguez, & Ostrand-Rosenberg, 2010). In models of lung

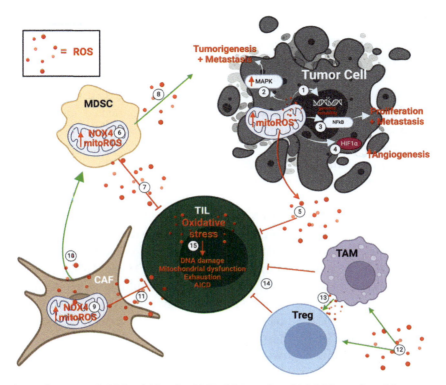

Fig. 1 **Sources of ROS within the TME.** (1) Mitochondrial ROS produced by tumor cells induces double stranded DNA breaks and leads to genomic instability. (2) ROS induces MAPK signaling to support tumorigenesis and metastasis. (3) ROS induces NFkβ signaling to increase expression of genes involved in proliferation and metastasis. (4) ROS stabilizes HIF1a expression to support angiogenesis. (5) ROS produced by tumor cells induces oxidative stress in T cells. (6) NOX4-derived ROS production in MDSCs. (7) ROS produced by MDSCs induces oxidative stress in T cells. (8) ROS produced by MDSCs stimulates tumorigenesis and metastasis. (9) NOX4-derived ROS production in CAFs. (10) ROS produced by CAFs stimulates MDSCs. (11) ROS produced by CAFs induces oxidative stress in T cells. (12) Increased ROS in the TME stimulates TAMs and Tregs. (13) TAM-derived ROS stimulates Treg differentiation. (14) TAMs and Tregs inhibit antitumor T cell activity. (15) Oxidative stress in TILs leads to DNA damage, mitochondrial dysfunction, exhaustion, AICD, and overall loss of antitumor activity.

squamous cell carcinoma, it was shown that inhibition of NOX2- and NOX4-dependent ROS generation by CAF-induced MDSCs restored CD8+ T cell function (Xiang et al., 2020).

Tumor associated macrophages (TAM) and Tregs: While not primary producers of ROS, many studies have demonstrated that elevated levels of ROS in the TME promote the differentiation and immunosuppressive

Table 1 Strategies targeting antioxidant pathway for improving immune response.

Target	Strategy	Impact	References
L-NAC/ glutathione	Expanding tumor-antigen specific T cells with L-NAC prior to adoptive transfer into melanoma tumor-bearing mice.	Improved in vivo persistence of adoptive transferred T cells, superior tumor control, and prolonged survival.	Scheffel, Scurti, and Simms (2016), Scheffel, Scurti, and Wyatt (2018)
Thiols	Treating tumor-antigen specific T cells with rapamycin or IL15 to induce thiol expression or sorting tumor-antigen specific T cells based on high expression of surface thiols prior to adoptive transfer.	Improved in vivo persistence of both rapamycin/IL15 treated T cells and thiol-high T cells, with superior tumor control and prolonged survival in all strategies.	Kesarwani et al. (2014), Kaur, Naga, and Norell (2011)
Thioredoxin	Expanding tumor-antigen specific T cells with recombinant thioredoxin (rTrx1) prior to adoptive transfer into melanoma tumor-bearing mice.	Improved in vivo persistence of adoptive transferred T cells, superior tumor control, and prolonged survival.	Chakraborty et al. (2019)
NAD+	Boosting intracellular NAD+ levels by differentiating Th1/Th17 hybrid T cells during expansion or blocking CD38 expression after transfer with an anti-CD38 antibody.	Improved in vivo persistence of T cells following both Th1/Th17 differentiation prior to transfer and administration of anti-CD38 antibody after transfer, with superior tumor control and prolonged survival in both strategies.	Chatterjee et al. (2018)

(continued)

Table 1 Strategies targeting antioxidant pathway for improving immune response. (*cont'd*)

Target	Strategy	Impact	References
Catalase	Co-expressing catalase in CAR–T (CAR-CAT-T) cells or treat *ex vivo* with recombinant catalase.	Decreased exhaustion and preserved T_{cm} phenotype in catalase treated T cells and CAR-CAT-T cells, with superior killing of target cells and bystander protection of non-transfect T and NK cells.	Ando et al. (2008), Ligtenberg et al. (2016), Aksoylar and Patsoukis (2023)
SOD	Treating human melanoma TILs with SOD mimetic MnTBAP to inhibit ROS.	Decreased susceptibility to AICD with preservation of effector function in human melanoma TILs.	Norell et al. (2009)
Gpx4	Overexpressing Gpx4 in TILs isolated from murine melanoma tumors prior to adoptive transfer.	Resolution of lipid peroxidation and restoration of effector function in murine melanoma TILs, with superior tumor control upon adoptive transfer.	Xu et al. (2021a)
ROS nanoscaven-gers	Directly inhibiting intratumoral ROS using a tumor extracellular matrix targeting ROS nanoscavenger in a murine breast cancer model.	Enhanced anti-tumor immune response in mice treated with the ROS nanoscavenger with increased infiltration of T cells and delayed tumor growth.	Deng, Yang, and Zhou (2020)

functions of TAMs and Tregs, thus contributing further to the immuno-suppressive nature of the TME (Christofides et al., 2022). Hydrogen peroxide secretion by TAMs can limit the cytotoxicity of NK cells and T cells (Kono, Salazar-Onfray, & Petersson, 1996). There is even evidence that TAMs can produce ROS to induce formation of Tregs, and that inhibition of NOX in macrophages limits Treg differentiation and regulatory function (Kraaij, Savage, & van der Kooij, 2010).

Collectively, these studies demonstrate the unique cellular milieu of TME that contributes to elevated ROS levels resulting in a highly immunosuppressive microenvironment (Fig. 1).

1.6 Antioxidant molecules and metabolites targeted in anti-tumor immunity

Several therapeutic strategies have been employed to boost antioxidant capacity in T cells to combat the high levels of ROS in the TME. A summary of the most promising strategies is outlined below in Table 1.

1.7 Clinical trials targeting immune cell antioxidant metabolism

While efforts have been made to assess the clinical effects of antioxidants taken as supplements for anti-cancer therapy, few clinical studies have been designed that specifically target the redox metabolism of immune cells to promote anti-tumor immunity. Phase I-III clinical trial in squamous cell carcinoma focused on supplementing patients with GC4419, a SOD mimic, in combination with chemoradiation to reduce side effects of the radiation on healthy tissues (NCT03689712). Our group's previous publications that demonstrated increased intracellular NAD+ levels in hybrid Th1/Th17 cells prolongs the anti-tumor response has led us to initiate a Phase 1b trial utilizing hybrid Th1/Th17 CD19 CAR-T cells in patients with relapsed or refract non-Hodgkin lymphoma or chronic lymphocytic leukemia (NCT05702853) (Chatterjee et al., 2018). In addition to our work, a Phase 1 trial currently enrolling is studying the effects of co-infusing N-acetylcysteine with CD19 CAR-T cells on the metabolism in the tumor microenvironment for patient with lymphoma (NCT05081479). These studies are an excellent starting point, but further efforts should be made to explore the clinical effects of modulating redox metabolism for cancer immunotherapy.

1.8 Conclusions and future directions

The tumor microenvironment is remarkably complex and dynamic. Tumor cells have evolved a host of immunosuppressive mechanisms that allow them to evade the adoptive immune response and grow unchecked. Thus, despite the advances we have made in understanding cancer immunology and designing therapies to direct the adaptive immune system to combat cancer, significant limitations remain in overall efficacy of cancer immunotherapy, particularly for solid tumors. One of the main contributors to immune cell dysregulation in the TME is the presence of a high level of reactive oxygen species that leads to oxidative stress of infiltrating T cells. Understanding redox dynamics within T cells will be critical in the pursuit to develop novel therapies that target oxidative stress in the TME, and this review highlights the preclinical strategies that have shown promise in targeting redox metabolism to improve immunotherapy outcomes. Further studies will be needed to push these strategies from preclinical models to more clinical trials.

Acknowledgment

The work was supported in part by NIH grants R01CA250458, R01CA236379, R41CA271958, R42CA239952 to SM, Support from South Caroline COBRE in Oxidants, Redox Balance and Stress Signaling Grant 5P30GM140964-02, Shared Instrumentation Grant (S10OD018113), and Hollings Cancer Center Shared Resources (partly supported by P30CA138313) at MUSC is also acknowledged.

References

Aksoylar, H.-I., & Patsoukis, N. (2023). Treatment with exogenously added catalase alters CD8 T cell memory differentiation and function. *Advanced Biology, 7*(4), 2101320. https://dci.org/10.1002/adbi.202101320.

Anderson, K. G., Stromnes, I. M., & Greenberg, P. D. (2017). Obstacles posed by the tumor microenvironment to T cell Activity: A case for synergistic therapies. *Cancer Cell, 31*(3), 311–325. https://doi.org/10.1016/j.ccell.2017.02.008.

Ando, T., Mimura, K., Johansson, C. C., et al. (2008). Transduction with the antioxidant enzyme catalase protects human T cells against oxidative stress. *Journal of Immunology, 181*(12), 8382–8390. https://doi.org/10.4049/jimmunol.181.12.8382.

Aquilano, K., Baldelli, S., & Ciriolo, M. R. (2014). Glutathione: New roles in redox signaling for an old antioxidant. Review. *Frontiers in Pharmacology, 5*. https://doi.org/10.3389/fphar.2014.00196.

Balliet, R. M., Capparelli, C., Guido, C., et al. (2011). Mitochondrial oxidative stress in cancer-associated fibroblasts drives lactate production, promoting breast cancer tumor growth. *Cell Cycle (Georgetown, Tex.), 10*(23), 4065–4073. https://doi.org/10.4161/cc.10.23.18254.

Battaglia, M., Stabilini, A., Migliavacca, B., Horejs-Hoeck, J., Kaupper, T., & Roncarolo, M. G. (2006). Rapamycin promotes expansion of functional CD4+CD25+FOXP3+ regulatory T cells of both healthy subjects and type 1 diabetic patients. *Journal of Immunology, 177*(12), 8338–8347. https://doi.org/10.4049/jimmunol.177.12.8338.

Bauer, M. K., Vogt, M., Los, M., Siegel, J., Wesselborg, S., & Schulze-Osthoff, K. (1998). Role of reactive oxygen intermediates in activation-induced CD95 (APO-1/Fas) ligand expression. *The Journal of Biological Chemistry, 273*(14), 8048–8055. https://doi.org/10.1074/jbc.273.14.8048.

Bhattacharyya, S., Mandal, D., Sen, G. S., et al. (2007). Tumor-induced oxidative stress perturbs nuclear factor-κB activity-augmenting tumor necrosis factor-α–mediated T-cell death: Protection by curcumin. *Cancer Research, 67*(1), 362–370. https://doi.org/10.1158/0008-5472.CAN-06-2583.

Cao, K., Wang, G., Li, W., et al. (2015). Histone deacetylase inhibitors prevent activation-induced cell death and promote anti-tumor immunity. *Oncogene, 34*(49), 5960–5970. https://doi.org/10.1038/onc.2015.46.

Case, A. J., McGill, J. L., Tygrett, L. T., et al. (2011). Elevated mitochondrial superoxide disrupts normal T cell development, impairing adaptive immune responses to an influenza challenge. *Free Radical Biology & Medicine, 50*(3), 448–458. https://doi.org/10.1016/j.freeradbiomed.2010.11.025.

Cemerski, S., Cantagrel, A., van Meerwijk, J. P. M., & Romagnoli, P. (2002). Reactive oxygen species differentially affect T cell receptor-signaling pathways*. *Journal of Biological Chemistry, 277*(22), 19585–19593. https://doi.org/10.1074/jbc.M111451200.

Chakraborty, P., Chatterjee, S., Kesarwani, P., et al. (2019). Thioredoxin-1 improves the immunometabolic phenotype of antitumor T cells. *The Journal of Biological Chemistry, 294*(23), 9198–9212. https://doi.org/10.1074/jbc.RA118.006753.

Chandel, N. S., Maltepe, E., Goldwasser, E., Mathieu, C. E., Simon, M. C., & Schumacker, P. T. (1998). Mitochondrial reactive oxygen species trigger hypoxia-induced transcription. *Proceedings of the National Academy of Sciences of the United States of America, 95*(20), 11715–11720. https://doi.org/10.1073/pnas.95.20.11715.

Chatterjee, S., Daenthanasanmak, A., Chakraborty, P., et al. (2018). CD38-NAD(+)axis regulates immunotherapeutic anti-tumor T cell response. *Cell Metabolism, 27*(1), 85–100.e8. https://doi.org/10.1016/j.cmet.2017.10.006.

Cheever, M. A., Greenberg, P. D., Irle, C., et al. (1984). Interleukin 2 administered in vivo induces the growth of cultured T cells in vivo. *Journal of Immunology, 132*(5), 2259–2265.

Chen, J., Liu, C., Chernatynskaya, A. V., et al. (2024). NADPH oxidase 2-derived reactive oxygen species promote CD8+ T cell effector function. *Journal of Immunology, 212*(2), 258–270. https://doi.org/10.4049/jimmunol.2200691.

Chen, W., Teo, J. M. N., Yau, S. W., et al. (2022). Chronic type I interferon signaling promotes lipid-peroxidation-driven terminal CD8(+) T cell exhaustion and curtails anti-PD-1 efficacy. *Cell Reports, 41*(7), 111647. https://doi.org/10.1016/j.celrep.2022.111647.

Chouchani, E. T., Pell, V. R., Gaude, E., et al. (2014). Ischaemic accumulation of succinate controls reperfusion injury through mitochondrial ROS. *Nature, 515*(7527), 431–435. https://doi.org/10.1038/nature13909.

Christofides, A., Strauss, L., Yeo, A., Cao, C., Charest, A., & Boussiotis, V. A. (2022). The complex role of tumor-infiltrating macrophages. *Nature Immunology, 23*(8), 1148–1156. https://doi.org/10.1038/s41590-022-01267-2.

Clay, T. M., Custer, M. C., Sachs, J., Hwu, P., Rosenberg, S. A., & Nishimura, M. I. (1999). Efficient transfer of a tumor antigen-reactive TCR to human peripheral blood lymphocytes confers anti-tumor reactivity. *Journal of Immunology, 163*(1), 507–513.

Corzo, C. A., Cotter, M. J., Cheng, P., et al. (2009). Mechanism regulating reactive oxygen species in tumor-induced myeloid-derived suppressor cells. *Journal of Immunology, 182*(9), 5693–5701. https://doi.org/10.4049/jimmunol.0900092.

Deng, H., Yang, W., Zhou, Z., et al. (2020). Targeted scavenging of extracellular ROS relieves suppressive immunogenic cell death. *Nature Communications, 11*(1), 4951. https://doi.org/10.1038/s41467-020-18745-6.

Devadas, S., Zaritskaya, L., Rhee, S. G., Oberley, L., & Williams, M. S. (2002). Discrete generation of superoxide and hydrogen peroxide by T cell receptor stimulation: Selective regulation of mitogen-activated protein kinase activation and fas ligand expression. *The Journal of Experimental Medicine, 195*(1), 59–70. https://doi.org/10.1084/jem.20010659.

Dolcet, X., Llobet, D., Pallares, J., & Matias-Guiu, X. (2005). NF-kB in development and progression of human cancer. *Virchows Archiv, 446*(5), 475–482. https://doi.org/10.1007/s00428-005-1264-9.

Duval, L., Schmidt, H., Kaltoft, K., et al. (2006). Adoptive transfer of allogeneic cytotoxic T lymphocytes equipped with a HLA-A2 restricted MART-1 T-cell receptor: A phase I trial in metastatic melanoma. *Clinical Cancer Research: An Official Journal of the American Association for Cancer Research, 12*(4), 1229–1236. https://doi.org/10.1158/1078-0432.Ccr-05-1485.

Ezernitchi, A. V., Vaknin, I., Cohen-Daniel, L., et al. (2006). TCR zeta down-regulation under chronic inflammation is mediated by myeloid suppressor cells differentially distributed between various lymphatic organs. *Journal of Immunology, 177*(7), 4763–4772. https://doi.org/10.4049/jimmunol.177.7.4763.

Flescher, E., Ledbetter, J. A., Schieven, G. L., et al. (1994). Longitudinal exposure of human T lymphocytes to weak oxidative stress suppresses transmembrane and nuclear signal transduction. *Journal of Immunology, 153*(11), 4880–4889.

Flescher, E., Tripoli, H., Salnikow, K., & Burns, F. J. (1998). Oxidative stress suppresses transcription factor activities in stimulated lymphocytes. *Clinical and Experimental Immunology, 112*(2), 242–247. https://doi.org/10.1046/j.1365-2249.1998.00548.x.

Frauwirth, K. A., Riley, J. L., Harris, M. H., et al. (2002). The CD28 signaling pathway regulates glucose metabolism. *Immunity, 16*(6), 769–777. https://doi.org/10.1016/S1074-7613(02)00323-0.

Frossi, B., De Carli, M., Piemonte, M., & Pucillo, C. (2008). Oxidative microenvironment exerts an opposite regulatory effect on cytokine production by Th1 and Th2 cells. *Molecular Immunology, 45*(1), 58–64. https://doi.org/10.1016/j.molimm.2007.05.008.

Fukai, T., & Ushio-Fukai, M. (2011). Superoxide dismutases: Role in redox signaling, vascular function, and diseases. *Antioxidants & Redox Signaling, 15*(6), 1583–1606. https://doi.org/10.1089/ars.2011.3999.

Gerriets, V. A., Kishton, R. J., Nichols, A. G., et al. (2015). Metabolic programming and PDHK1 control CD4+ T cell subsets and inflammation. *The Journal of Clinical Investigation, 125*(1), 194–207. https://doi.org/10.1172/jci76012.

Greenberg, P. D., Cheever, M. A., & Fefer, A. (1981). Eradication of disseminated murine leukemia by chemoimmunotherapy with cyclophosphamide and adoptively transferred immune syngeneic Lyt-1+2- lymphocytes. *The Journal of Experimental Medicine, 154*(3), 952–963. https://doi.org/10.1084/jem.154.3.952.

Hahne, M., Rimoldi, D., Schröter, M., et al. (1996). Melanoma cell expression of Fas(Apo-1/CD95) ligand: Implications for tumor immune escape. *Science (New York, N. Y.), 274*(5291), 1363–1366. https://doi.org/10.1126/science.274.5291.1363.

Hanley, C. J., Mellone, M., Ford, K., et al. (2018). Targeting the myofibroblastic cancer-associated fibroblast phenotype through inhibition of NOX4. *JNCI: Journal of the National Cancer Institute, 110*(1), 109–120. https://doi.org/10.1093/jnci/djx121.

Hart, P. C., Mao, M., de Abreu, A. L., et al. (2015). MnSOD upregulation sustains the Warburg effect via mitochondrial ROS and AMPK-dependent signalling in cancer. *Nature Communications, 6*, 6053. https://doi.org/10.1038/ncomms7053.

Hildeman, D. A., Mitchell, T., Kappler, J., & Marrack, P. (2003). T cell apoptosis and reactive oxygen species. *The Journal of Clinical Investigation, 111*(5), 575–581. https://doi.org/10.1172/JCI18007.

Ippolito, L., Morandi, A., Taddei, M. L., et al. (2019). Cancer-associated fibroblasts promote prostate cancer malignancy via metabolic rewiring and mitochondrial transfer. *Oncogene, 38*(27), 5339–5355. https://doi.org/10.1038/s41388-019-0805-7.

Iwata, S., Hori, T., Sato, N., et al. (1997). Adult T cell leukemia (ATL)-derived factor/human thioredoxin prevents apoptosis of lymphoid cells induced by L-cystine and glutathione depletion: Possible involvement of thiol-mediated redox regulation in apoptosis caused by pro-oxidant state. *Journal of Immunology, 158*(7), 3108–3117.

Jackson, S. H., Devadas, S., Kwon, J., Pinto, L. A., & Williams, M. S. (2004). T cells express a phagocyte-type NADPH oxidase that is activated after T cell receptor stimulation. *Nature Immunology, 5*(8), 818–827. https://doi.org/10.1038/ni1096.

Jeggo, P. A., & Löbrich, M. (2007). DNA double-strand breaks: Their cellular and clinical impact? *Oncogene, 26*(56), 7717–7719. https://doi.org/10.1038/sj.onc.1210868.

Johnson, W. M., Wilson-Delfosse, A. L., & Mieyal, J. J. (2012). Dysregulation of glutathione homeostasis in neurodegenerative diseases. *Nutrients, 4*(10), 1399–1440. https://doi.org/10.3390/nu4101399.

Kamiński, M. M., Röth, D., Sass, S., Sauer, S. W., Krammer, P. H., & Gülow, K. (2012b). Manganese superoxide dismutase: A regulator of T cell activation-induced oxidative signaling and cell death. *Biochimica et Biophysica Acta, 1823*(5), 1041–1052. https://doi.org/10.1016/j.bbamcr.2012.03.003.

Kamiński, M. M., Sauer, S. W., Kamiński, M., et al. (2012a). T cell activation is driven by an ADP-dependent glucokinase linking enhanced glycolysis with mitochondrial reactive oxygen species generation. *Cell Reports, 2*(5), 1300–1315. https://doi.org/10.1016/j.celrep.2012.10.009.

Kaminski, M. M., Sauer, S. W., Klemke, C. D., et al. (2010). Mitochondrial reactive oxygen species control T cell activation by regulating IL-2 and IL-4 expression: Mechanism of ciprofloxacin-mediated immunosuppression. *Journal of Immunology, 184*(9), 4827–4841. https://doi.org/10.4049/jimmunol.0901662.

Kaur, N., Naga, O. S., Norell, H., et al. (2011). T cells expanded in presence of IL-15 exhibit increased antioxidant capacity and innate effector molecules. *Cytokine, 55*(2), 307–317. https://doi.org/10.1016/j.cyto.2011.04.014.

Kawakami, Y., & Rosenberg, S. A. (1996). T-cell recognition of self peptides as tumor rejection antigens. *Immunologic Research, 15*(3), 179–190. https://doi.org/10.1007/bf02918248.

Kesarwani, P., Al-Khami, A. A., Scurti, G., et al. (2014). Promoting thiol expression increases the durability of antitumor T-cell functions. *Cancer Research, 74*(21), 6036–6047. https://doi.org/10.1158/0008-5472.Can-14-1084.

Kim, H. J., Barajas, B., Wang, M., & Nel, A. E. (2008). Nrf2 activation by sulforaphane restores the age-related decrease of T(H)1 immunity: Role of dendritic cells. *The Journal of Allergy and Clinical Immunology, 121*(5), 1255–1261.e7. https://doi.org/10.1016/j.jaci.2008.01.016.

Kirtonia, A., Sethi, G., & Garg, M. (2020). The multifaceted role of reactive oxygen species in tumorigenesis. *Cellular and Molecular Life Sciences, 77*(22), 4459–4483. https://doi.org/10.1007/s00018-020-03536-5.

Kono, K., Salazar-Onfray, F., Petersson, M., et al. (1996). Hydrogen peroxide secreted by tumor-derived macrophages down-modulates signal-transducing zeta molecules and inhibits tumor-specific T cell-and natural killer cell-mediated cytotoxicity. *European Journal of Immunology, 26*(6), 1308–1313. https://doi.org/10.1002/eji.1830260620.

Kraaij, M. D., Savage, N. D., van der Kooij, S. W., et al. (2010). Induction of regulatory T cells by macrophages is dependent on production of reactive oxygen species. *Proceedings of the National Academy of Sciences of the United States of America, 107*(41), 17686–17691. https://doi.org/10.1073/pnas.1012016107.

Lee, S., Kim, S. M., & Lee, R. T. (2013). Thioredoxin and thioredoxin target proteins: From molecular mechanisms to functional significance. *Antioxidants & Redox Signaling, 18*(10), 1165–1207. https://doi.org/10.1089/ars.2011.4322.

Lee, D. H., Son, D. J., Park, M. H., Yoon, D. Y., Han, S. B., & Hong, J. T. (2016). Glutathione peroxidase 1 deficiency attenuates concanavalin A-induced hepatic injury by modulation of T-cell activation. *Cell Death & Disease, 7*(4), e2208. https://doi.org/10.1038/cddis.2016.95.

Lee, S.-R., Yang, K.-S., Kwon, J., Lee, C., Jeong, W., & Rhee, S. G. (2002). Reversible inactivation of the tumor suppressor PTEN by H2O2*. *Journal of Biological Chemistry, 277*(23), 20336–20342. https://doi.org/10.1074/jbc.M111899200.

Ligtenberg, M. A., Mougiakakos, D., Mukhopadhyay, M., et al. (2016). Coexpressed catalase protects chimeric antigen receptor-redirected T cells as well as bystander cells from oxidative stress-induced loss of antitumor activity. *Journal of Immunology, 196*(2), 759–766. https://doi.org/10.4049/jimmunol.1401710.

Li, Q., & Engelhardt, J. F. (2006). Interleukin-1beta induction of NFkappaB is partially regulated by H2O2-mediated activation of NFkappaB-inducing kinase. *The Journal of Biological Chemistry, 281*(3), 1495–1505. https://doi.org/10.1074/jbc.M511153200.

Martinez-Outschoorn, U. E., Balliet, R. M., Rivadeneira, D. B., et al. (2010). Oxidative stress in cancer associated fibroblasts drives tumor-stroma co-evolution: A new paradigm for understanding tumor metabolism, the field effect and genomic instability in cancer cells. *Cell Cycle (Georgetown, Tex.), 9*(16), 3256–3276. https://doi.org/10.4161/cc.9.16.12553.

Matsushita, M., Freigang, S., Schneider, C., Conrad, M., Bornkamm, G. W., & Kopf, M. (2015). T cell lipid peroxidation induces ferroptosis and prevents immunity to infection. *Journal of Experimental Medicine, 212*(4), 555–568. https://doi.org/10.1084/jem.20140857.

Mazzoni, A., Bronte, V., Visintin, A., et al. (2002). Myeloid suppressor lines inhibit T cell responses by an NO-dependent mechanism. *Journal of Immunology, 168*(2), 689–695. https://doi.org/10.4049/jimmunol.168.2.689.

Mehrotra, S., Chhabra, A., Chattopadhyay, S., Dorsky, D. I., Chakraborty, N. G., & Mukherji, B. (2004). Rescuing melanoma epitope-specific cytolytic T lymphocytes from activation-induced cell death, by SP600125, an inhibitor of JNK: Implications in cancer immunotherapy. *Journal of Immunology, 173*(10), 6017–6024. https://doi.org/10.4049/jimmunol.173.10.6017.

Messina, S., De Simone, G., & Ascenzi, P. (2019). Cysteine-based regulation of redox-sensitive Ras small GTPases. *Redox Biology, 26*, 101282. https://doi.org/10.1016/j.redox.2019.101282.

Meves, A., Stock, S. N., Beyerle, A., Pittelkow, M. R., & Peus, D. (2001). H(2)O(2) mediates oxidative stress-induced epidermal growth factor receptor phosphorylation. *Toxicology Letters, 122*(3), 205–214. https://doi.org/10.1016/s0378-4274(01)00359-9.

Morandi, F., Horenstein, A. L., & Malavasi, F. (2021). The key role of NAD+ in anti-tumor immune response: An update. Mini review. *Frontiers in Immunology, 12*. https://doi.org/10.3389/fimmu.2021.658263.

Morgan, R. A., Dudley, M. E., Wunderlich, J. R., et al. (2006). Cancer regression in patients after transfer of genetically engineered lymphocytes. *Science (New York, N. Y.), 314*(5796), 126–129. https://doi.org/10.1126/science.1129003.

Morgan, M. J., & Liu, Z.-G. (2011). Crosstalk of reactive oxygen species and NF-κB signaling. *Cell Research, 21*(1), 103–115. https://doi.org/10.1038/cr.2010.178.

Morito, N., Yoh, K., Itoh, K., et al. (2003). Nrf2 regulates the sensitivity of death receptor signals by affecting intracellular glutathione levels. *Oncogene, 22*(58), 9275–9281. https://doi.org/10.1038/sj.onc.1207024.

Morzadec, C., Macoch, M., Sparfel, L., Kerdine-Römer, S., Fardel, O., & Vernhet, L. (2014). Nrf2 expression and activity in human T lymphocytes: Stimulation by T cell receptor activation and priming by inorganic arsenic and tert-butylhydroquinone. *Free Radical Biology & Medicine, 71*, 133–145. https://doi.org/10.1016/j.freeradbiomed.2014.03.006.

Mougiakakos, D., Johansson, C. C., Jitschin, R., Böttcher, M., & Kiessling, R. (2011). Increased thioredoxin-1 production in human naturally occurring regulatory T cells confers enhanced tolerance to oxidative stress. *Blood, 117*(3), 857–861. https://doi.org/10.1182/blood-2010-09-307041.

Mougiakakos, D., Johansson, C. C., & Kiessling, R. (2009). Naturally occurring regulatory T cells show reduced sensitivity toward oxidative stress-induced cell death. *Blood, 113*(15), 3542–3545. https://doi.org/10.1182/blood-2008-09-181040.

Muller, F. L., Liu, Y., & Van Remmen, H. (2004). Complex III releases superoxide to both sides of the inner mitochondrial membrane. *The Journal of Biological Chemistry, 279*(47), 49064–49073. https://doi.org/10.1074/jbc.M407715200.

Nagaraj, S., Gupta, K., Pisarev, V., et al. (2007). Altered recognition of antigen is a mechanism of CD8+ T cell tolerance in cancer. *Nature Medicine, 13*(7), 828–835. https://doi.org/10.1038/nm1609.

Nakamura, T., Oh, C.-k, Zhang, X., & Lipton, S. A. (2021). Protein S-nitrosylation and oxidation contribute to protein misfolding in neurodegeneration. *Free Radical Biology and Medicine, 172*, 562–577. https://doi.org/10.1016/j.freeradbiomed.2021.07.002.

Norell, H., Martins da Palma, T., Lesher, A., et al. (2009). Inhibition of superoxide generation upon T-cell receptor engagement rescues Mart-1(27-35)-reactive T cells from activation-induced cell death. *Cancer Research, 69*(15), 6282–6289. https://doi.org/10.1158/0008-5472.Can-09-1176.

O'Brien, M., Chalker, J., Slade, L., Gardiner, D., & Mailloux, R. J. (2017). Protein S-glutathionylation alters superoxide/hydrogen peroxide emission from pyruvate dehydrogenase complex. *Free Radical Biology & Medicine, 106*, 302–314. https://doi.org/10.1016/j.freeradbiomed.2017.02.046.

Old, L. J. (1982). Cancer immunology: The search for specificity. *National Cancer Institute Monograph, 60*, 193–209.

Orimo, A., Gupta, P. B., Sgroi, D. C., et al. (2005). Stromal fibroblasts present in invasive human breast carcinomas promote tumor growth and angiogenesis through elevated SDF-1/CXCL12 secretion. *Cell, 121*(3), 335–348. https://doi.org/10.1016/j.cell.2005.02.034.

Orr, A. L., Vargas, L., Turk, C. N., et al. (2015). Suppressors of superoxide production from mitochondrial complex III. *Nature Chemical Biology, 11*(11), 834–836. https://doi.org/10.1038/nchembio.1910.

Ostrand-Rosenberg, S., Sinha, P., Beury, D. W., & Clements, V. K. (2012). Cross-talk between myeloid-derived suppressor cells (MDSC), macrophages, and dendritic cells enhances tumor-induced immune suppression. *Seminars in Cancer Biology, 22*(4), 275–281. https://doi.org/10.1016/j.semcancer.2012.01.011.

Otsuji, M., Kimura, Y., Aoe, T., Okamoto, Y., & Saito, T. (1996). Oxidative stress by tumor-derived macrophages suppresses the expression of CD3 zeta chain of T-cell receptor complex and antigen-specific T-cell responses. *Proceedings of the National Academy of Sciences of the United States of America, 93*(23), 13119–13124. https://doi.org/10.1073/pnas.93.23.13119.

Pardo, M., Melendez, J. A., & Tirosh, O. (2006). Manganese superoxide dismutase inactivation during Fas (CD95)-mediated apoptosis in Jurkat T cells. *Free Radical Biology & Medicine, 41*(12), 1795–1806. https://doi.org/10.1016/j.freeradbiomed.2006.08.027.

Park, K. R., Nam, D., Yun, H. M., et al. (2011). β-Caryophyllene oxide inhibits growth and induces apoptosis through the suppression of PI3K/AKT/mTOR/S6K1 pathways and ROS-mediated MAPKs activation. *Cancer Letters, 312*(2), 178–188. https://doi.org/10.1016/j.canlet.2011.08.001.

Pizzorno, J. (2014). Glutathione!. *Integrative Medicin (Encinitas), 13*(1), 8–12.

Pugh, C. W., & Ratcliffe, P. J. (2003). Regulation of angiogenesis by hypoxia: Role of the HIF system. *Nature Medicine, 9*(6), 677–684. https://doi.org/10.1038/nm0603-677.

Ray, P. D., Huang, B. W., & Tsuji, Y. (2012). Reactive oxygen species (ROS) homeostasis and redox regulation in cellular signaling. *Cellular Signalling, 24*(5), 981–990. https://doi.org/10.1016/j.cellsig.2012.01.008.

Rengarajan, J., Mittelstadt, P. R., Mages, H. W., et al. (2000). Sequential involvement of NFAT and Egr transcription factors in FasL regulation. *Immunity, 12*(3), 293–300. https://doi.org/10.1016/s1074-7613(00)80182-x.

Requejo, R., Hurd, T. R., Costa, N. J., & Murphy, M. P. (2010). Cysteine residues exposed on protein surfaces are the dominant intramitochondrial thiol and may protect against oxidative damage. *The FEBS Journal, 277*(6), 1465–1480. https://doi.org/10.1111/j.1742-4658.2010.07576.x.

Rice, C. M., Davies, L. C., Subleski, J. J., et al. (2018). Tumour-elicited neutrophils engage mitochondrial metabolism to circumvent nutrient limitations and maintain immune suppression. *Nature Communications, 9*(1), 5099. https://doi.org/10.1038/s41467-018-07505-2.

Ruiz-Ramos, R., Lopez-Carrillo, L., Rios-Perez, A. D., De Vizcaya-Ruíz, A., & Cebrian, M. E. (2009). Sodium arsenite induces ROS generation, DNA oxidative damage, HO-1 and c-Myc proteins, NF-kappaB activation and cell proliferation in human breast cancer MCF-7 cells. *Mutation Research, 674*(1–2), 109–115. https://doi.org/10.1016/j.mrgentox.2008.09.021.

Salmeen, A., Andersen, J. N., Myers, M. P., et al. (2003). Redox regulation of protein tyrosine phosphatase 1B involves a sulphenyl-amide intermediate. *Nature, 423*(6941), 769–773. https://doi.org/10.1038/nature01680.

Sandstrom, P. A., & Buttke, T. M. (1993). Autocrine production of extracellular catalase prevents apoptosis of the human CEM T-cell line in serum-free medium. *Proceedings of the National Academy of Sciences of the United States of America, 90*(10), 4708–4712. https://doi.org/10.1073/pnas.90.10.4708.

Scheffel, M. J., Scurti, G., Simms, P., et al. (2016). Efficacy of adoptive T-cell therapy is improved by treatment with the antioxidant N-acetyl cysteine, which limits activation-induced T-cell death. *Cancer Research, 76*(20), 6006–6016. https://doi.org/10.1158/0008-5472.CAN-16-0587.

Scheffel, M. J., Scurti, G., Wyatt, M. M., et al. (2018). N-acetyl cysteine protects anti-melanoma cytotoxic T cells from exhaustion induced by rapid expansion via the downmodulation of Foxo1 in an Akt-dependent manner. *Cancer Immunology, Immunotherapy: CII, 67*(4), 691–702. https://doi.org/10.1007/s00262-018-2120-5.

Schieven, G. L., Mittler, R. S., Nadler, S. G., et al. (1994). ZAP-70 tyrosine kinase, CD45, and T cell receptor involvement in UV- and H2O2-induced T cell signal transduction. *Journal of Biological Chemistry, 269*(32), 20718–20726.

Sena, L. A., Li, S., Jairaman, A., et al. (2013a). Mitochondria are required for antigen-specific T cell activation through reactive oxygen species signaling. *Immunity, 38*(2), 225–236. https://doi.org/10.1016/j.immuni.2012.10.020.

Sena, L. A., Li, S., Jairaman, A., et al. (2013b). Mitochondria are required for antigen-specific T cell activation through reactive oxygen species signaling. *Immunity, 38*(2), 225–236. https://doi.org/10.1016/j.immuni.2012.10.020.

Shields, H. J., Traa, A., & Van Raamsdonk, J. M. (2021). Beneficial and detrimental effects of reactive oxygen species on lifespan: A comprehensive review of comparative and experimental studies. *Frontiers in Cell and Developmental Biology, 9*, 628157. https://doi.org/10.3389/fcell.2021.628157.

Shin, S.-Y., Kim, M.-W., Cho, K.-H., & Nguyen, L. K. (2019). Coupled feedback regulation of nuclear factor of activated T-cells (NFAT) modulates activation-induced cell death of T cells. *Scientific Reports, 9*(1), 10637. https://doi.org/10.1038/s41598-019-46592-z.

Someya, S., Yu, W., Hallows, W. C., et al. (2010). Sirt3 mediates reduction of oxidative damage and prevention of age-related hearing loss under caloric restriction. *Cell, 143*(5), 802–812. https://doi.org/10.1016/j.cell.2010.10.002.

Son, Y., Cheong, Y. K., Kim, N. H., Chung, H. T., Kang, D. G., & Pae, H. O. (2011). Mitogen-activated protein kinases and reactive oxygen species: How can ROS activate MAPK pathways? *Journal of Signal Transduction, 2011*, 792639. https://doi.org/10.1155/2011/792639.

Srinivas, U. S., Tan, B. W. Q., Vellayappan, B. A., & Jeyasekharan, A. D. (2019). ROS and the DNA damage response in cancer. *Redox Biology, 25*, 101084. https://doi.org/10.1016/j.redox.2018.101084.

Srivastava, M. K., Sinha, P., Clements, V. K., Rodriguez, P., & Ostrand-Rosenberg, S. (2010). Myeloid-derived suppressor cells inhibit T-cell activation by depleting cystine and cysteine. *Cancer Research, 70*(1), 68–77. https://doi.org/10.1158/0008-5472.Can-09-2587.

Sundaresan, N. R., Gupta, M., Kim, G., Rajamohan, S. B., Isbatan, A., & Gupta, M. P. (2009). Sirt3 blocks the cardiac hypertrophic response by augmenting Foxo3a-dependent antioxidant defense mechanisms in mice. *The Journal of Clinical Investigation, 119*(9), 2758–2771. https://doi.org/10.1172/jci39162.

Szatrowski, T. P., & Nathan, C. F. (1991). Production of large amounts of hydrogen peroxide by human tumor cells. *Cancer Research, 51*(3), 794–798.

Takahashi, A., Hanson, M. G., Norell, H. R., et al. (2005). Preferential cell death of CD8+ effector memory (CCR7-CD45RA-) T cells by hydrogen peroxide-induced oxidative stress. *Journal of Immunology, 174*(10), 6080–6087. https://doi.org/10.4049/jimmunol.174.10.6080.

Topalian, S. L., Hom, S. S., Kawakami, Y., et al. (1992). Recognition of shared melanoma antigens by human tumor-infiltrating lymphocytes. *Journal of Immunotherapy: Official Journal of the Society for Biological Therapy, 12*(3), 203–206. https://doi.org/10.1097/00002371-199210000-00013.

Tretter, L., & Adam-Vizi, V. (2004). Generation of reactive oxygen species in the reaction catalyzed by alpha-ketoglutarate dehydrogenase. *The Journal of Neuroscience, 24*(36), 7771–7778. https://doi.org/10.1523/jneurosci.1842-04.2004.

Tribble, D. L., Aw, T. Y., & Jones, D. P. (1987). The pathophysiological significance of lipid peroxidation in oxidative cell injury. *Hepatology (Baltimore, Md.), 7*(2).

Tse, H. M., Thayer, T. C., Steele, C., et al. (2010). NADPH oxidase deficiency regulates Th lineage commitment and modulates autoimmunity. *Journal of Immunology, 185*(9), 5247–5258. https://doi.org/10.4049/jimmunol.1001472.

Turrens, J. F. (2003). Mitochondrial formation of reactive oxygen species. *The Journal of Physiology, 552*(Pt 2), 335–344. https://doi.org/10.1113/jphysiol.2003.049478.

Ulrich, K., & Jakob, U. (2019). The role of thiols in antioxidant systems. *Free Radical Biology & Medicine, 140*, 14–27. https://doi.org/10.1016/j.freeradbiomed.2019.05.035.

Vardhana, S. A., Hwee, M. A., Berisa, M., et al. (2020). Impaired mitochondrial oxidative phosphorylation limits the self-renewal of T cells exposed to persistent antigen. *Nature Immunology, 21*(9), 1022–1033. https://doi.org/10.1038/s41590-020-0725-2.

Vomund, S., Schäfer, A., Parnham, M. J., Brüne, B., & von Knethen, A. (2017). Nrf2, the master regulator of anti-oxidative responses. *International Journal of Molecular Sciences, 18*(12), https://doi.org/10.3390/ijms18122772.

Wang, G. L., Jiang, B. H., Rue, E. A., & Semenza, G. L. (1995). Hypoxia-inducible factor 1 is a basic-helix-loop-helix-PAS heterodimer regulated by cellular O2 tension. *Proceedings of the National Academy of Sciences of the United States of America, 92*(12), 5510–5514. https://doi.org/10.1073/pnas.92.12.5510.

Wang, Y., Wang, F., Wang, L., et al. (2021). NAD(+) supplement potentiates tumor-killing function by rescuing defective TUB-mediated NAMPT transcription in tumor-infiltrated T cells. *Cell Reports, 36*(6), 109516. https://doi.org/10.1016/j.celrep.2021.109516.

Weinberg, F., Hamanaka, R., Wheaton, W. W., et al. (2010). Mitochondrial metabolism and ROS generation are essential for Kras-mediated tumorigenicity. *Proceedings of the National Academy of Sciences of the United States of America, 107*(19), 8788–8793. https://doi.org/10.1073/pnas.1003428107.

Weinberg, F., Ramnath, N., & Nagrath, D. (2019). Reactive oxygen species in the tumor microenvironment: An overview. *Cancers (Basel), 11*(8), https://doi.org/10.3390/cancers11081191.

Wilson, B. E., Mochon, E., & Boxer, L. M. (1996). Induction of bcl-2 expression by phosphorylated CREB proteins during B-cell activation and rescue from apoptosis. *Molecular and Cellular Biology, 16*(10), 5546–5556. https://doi.org/10.1128/mcb.16.10.5546.

Won, H. Y., Sohn, J. H., Min, H. J., et al. (2010). Glutathione peroxidase 1 deficiency attenuates allergen-induced airway inflammation by suppressing Th2 and Th17 cell development. *Antioxidants & Redox Signaling, 13*(5), 575–587. https://doi.org/10.1089/ars.2009.2989.

Xiang, H., Ramil, C. P., Hai, J., et al. (2020). Cancer-associated fibroblasts promote immunosuppression by inducing ROS-generating monocytic MDSCs in lung squamous cell carcinoma. *Cancer Immunology Research, 8*(4), 436–450. https://doi.org/10.1158/2326-6066.Cir-19-0507.

Xu, S., Chaudhary, O., Rodríguez-Morales, P., et al. (2021a). Uptake of oxidized lipids by the scavenger receptor CD36 promotes lipid peroxidation and dysfunction in CD8(+) T cells in tumors. *Immunity, 54*(7), 1561–1577.e7. https://doi.org/10.1016/j.immuni.2021.05.003.

Xu, C., Sun, S., Johnson, T., et al. (2021b). The glutathione peroxidase Gpx4 prevents lipid peroxidation and ferroptosis to sustain Treg cell activation and suppression of antitumor immunity. *Cell Reports, 35*(11), 109235. https://doi.org/10.1016/j.celrep.2021.109235.

Yang, L., DeBusk, L. M., Fukuda, K., et al. (2004). Expansion of myeloid immune suppressor Gr+CD11b+ cells in tumor-bearing host directly promotes tumor angiogenesis. *Cancer Cell, 6*(4), 409–421. https://doi.org/10.1016/j.ccr.2004.08.031.

Yu, Y.-R., Imrichova, H., Wang, H., et al. (2020). Disturbed mitochondrial dynamics in CD8+ TILs reinforce T cell exhaustion. *Nature Immunology, 21*(12), 1540–1551. https://doi.org/10.1038/s41590-020-0793-3.

Zhang, Y., Zhang, M., Zhu, W., et al. (2020). Succinate accumulation induces mitochondrial reactive oxygen species generation and promotes status epilepticus in the kainic acid rat model. *Redox Biology, 28*, 101365. https://doi.org/10.1016/j.redox.2019.101365.

Zhao, R. Z., Jiang, S., Zhang, L., & Yu, Z. B. (2019). Mitochondrial electron transport chain, ROS generation and uncoupling (Review). *International Journal of Molecular Medicine, 44*(1), 3–15. https://doi.org/10.3892/ijmm.2019.4188.

CHAPTER FIVE

Redox pathways in melanoma

Jie Zhang[a,*], Zhi-wei Ye[a], Danyelle M. Townsend[b], and Kenneth D. Tew[a]

[a]Department of Cell and Molecular Pharmacology and Experimental Therapeutics, Medical University of South Carolina, Charleston, SC, United States
[b]Department of Drug Discovery and Biomedical Sciences, Medical University of South Carolina, Charleston, SC, United States
*Corresponding author. e-mail address: zhajie@musc.edu

Contents

1. Introduction	126
2. Melanin biosynthesis	127
3. MGST1 and melanoma	131
4. MGST1 in melanoma metastasis and treatment response	132
5. Additional redox targets in melanoma	135
Acknowledgments	140
References	140

Abstract

Cases of melanoma are doubling every 12 years, and in stages III and IV, the disease is associated with high mortality rates concomitant with unresectable metastases and therapeutic drug resistance. Despite some advances in treatment success, there is a marked need to understand more about the pathology of the disease. The present review provides an overview of how melanoma cells use and modulate redox pathways to facilitate thiol homeostasis and melanin biosynthesis and describes plausible redox targets that may improve therapeutic approaches in managing malignant disease and metastasis. Melanotic melanoma has some unique characteristics. Making melanin requires a considerable dedication of cellular energy resources and utilizes glutathione and glutathione transferases in certain steps in the biosynthetic pathway. Melanin is an antioxidant but is also functionally important in hematopoiesis and influential in various aspects of host immune responses, giving it unique characteristics. Together with other redox traits that are specific to melanoma, a discussion of possible therapeutic approaches is also provided.

Abbreviations

ACSL3	acyl-CoA Synthetase Long Chain Family Member 3
BRAFi	BRAF inhibitors
ER	endoplasmic reticulum
GPX	glutathione-dependent peroxidase

GSH	glutathione
GST	glutathione transferases
HO-1	heme oxygenase 1
KD	knockdown
MAPEG	membrane-associated proteins
MGST1	microsomal glutathione transferase 1
NAC	N-acetylcysteine
PPP	pentose phosphate pathway
PRDX3	mitochondrial thioredoxin-dependent peroxide reductase
ROS	reactive oxygen species
SOD	superoxide dismutases
SREBP-1	sterol regulatory binding element 1
TR	thioredoxin reductase
TRX	thioredoxin
TXNIP	thioredoxin interacting protein
UM	uveal melanoma

1. Introduction

Melanoma is the deadliest form of skin cancer. Although prevention and early detection strategies, and the development of more effective treatments (targeted BRAF or MEK inhibitors, immunotherapy, and combinations thereof) have improved survival, many patients experience therapy resistance with consequent mortality. Moreover, even with these advances, metastatic melanoma (Stage IV) only has a 5-year survival rate of around 35%.[1]

Like many tumor types, melanoma cells maintain a generally aberrant redox homeostasis when compared to normal cells. Metastases from a primary melanoma can differ in their redox homeostatic parameters, despite originating from the genetically related (but heterogeneous) primary tumor. This variance can be influenced by several factors: amongst them, genetic drift, conditions at the metastatic site and the influence of surrounding tissue elements. Since disease cure is frequently contingent upon the elimination of metastatic lesions, understanding such differences is an essential consideration.

Oxidative stress plays a role in melanoma progression, promoting tumor initiation, while paradoxically hindering vertical growth and metastasis in later disease (Becker & Indra, 2023). Following tumor initiation, endogenous antioxidant systems are co-opted by adaptive metabolic reprogramming to reduce levels of oxidative stress. This altered redox homeostasis contributes

---

[1] Cancer Stat Facts: Melanoma of the Skin. https://seer.cancer.gov/statfacts/html/melan.html.

to tumor progression and metastasis, while also complicating the application of exogenous treatments(Carpenter, Becker, & Indra, 2022). Metabolic rewiring of redox pathways and redox homeostasis is implicated in acquired resistance to many drugs including, the initially effective BRAF/MEK inhibitors. In this regard, one suggested approach to enhance therapeutic success involves boosting intracellular reactive oxygen species (ROS) production using active biomolecules or targeting enzymes that regulate oxidative stress (Becker & Indra, 2023). The complex interplay between oxidative stress, redox homeostasis, and melanomagenesis can also be leveraged in a preventive context. The intersection of melanin and melanoma pathology is a characteristic unique to this disease.

This review aims to provide an overview of redox pathways in melanoma and their metastases. Melanin biosynthesis includes steps that use glutathione (GSH) and associated transferase enzymes. We include an analysis of the involvement of glutathione transferases (GST) in this process and how antioxidant systems may be plausible therapeutic targets in the context of improved efficacy and survival.

2. Melanin biosynthesis

Melanin is a skin-associated piment, with antioxidant properties. Melanin production depends upon redox conditions for some of the intermediate steps in its synthesis and melanotic melanomas are known to express wide ranges of intracellular melanin coloration. Abnormal accumulation of melanin is linked with some types of malignant melanoma and melanogenesis is associated with shortened overall survival and disease-free survival in patients with metastatic disease (Brozyna, Jozwicki, Carlson, & Slominski, 2013). In normal epidermal melanocytes melanin pigment serves to protect from a variety of insults by functioning as a light filter, radioprotector and scavenger of free radicals, metal cations, and many types of electrophilic chemicals. Under physiological conditions, these properties are beneficial to the skin and the evolutionary importance of these protective properties is underscored by the energy commitment of the cells to make the pigment. However, in cancers, melanin can contribute to chemo-, radio-, and phototherapy resistance. Further, melanogenesis generates an immunosuppressive and mutagenic environment and alters cellular metabolism (Slominski & Carlson, 2014; Slominski, Zmijewski, & Slominski, 2015). Because of the intrinsic ability of melanomas to develop

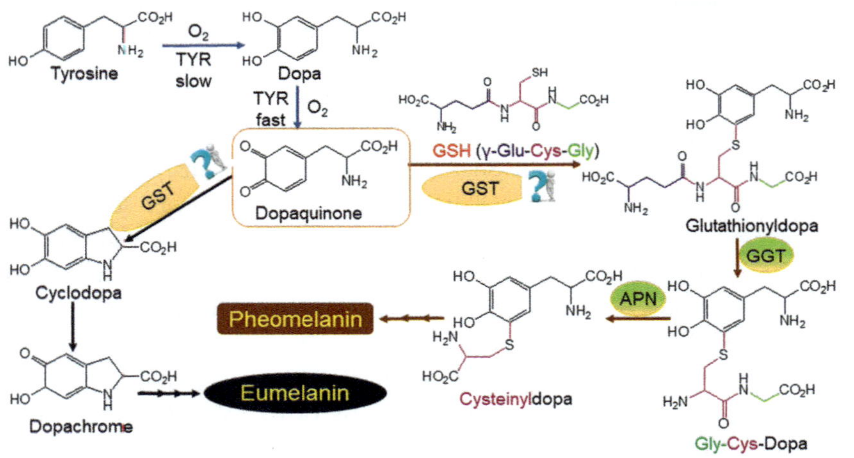

Fig. 1 GST/GSH involvement in melanin biosynthesis. Pheomelanin and eumelanin are synthesized from the amino acid tyrosine, with dopaquinone as an intermediate metabolite, catalyzed by tyrosinase (TYR). Quinones are electrophilic species and the formation of a GSH-dopaquinone conjugate, catalyzed by GSTP and GSTM2-2, serves both to "detoxify" and create an intermediate in melanin synthesis. Sequential in the biosynthesis, gamma-glutamyl transpeptidase (GGT) and aminopeptidase N (APN) remove two of the GSH amino acids to leave cysteinyldopa, a catecholamine. In addition to the pheomelanin pathway, dopaquinone can undergo slow intramolecular cyclization to cyclodopa, followed by rapid oxidation to dopachrome, the precursor of eumelanin. In the step shown, MGST1 may operate as a cyclase by stabilization of the deprotonated dopaquinone.

resistance to therapy, there is a need to identify molecular pathways and signatures associated with tumor progression. Consequently, melanogenesis may represent a valid therapeutic target for anti-melanoma treatments.

Fig. 1 shows those biochemical steps that underlie melanin biosynthesis, illustrating those steps that involve redox components, GST and GSH. This scheme highlights the present understanding of how pheomelanin and eumelanin are synthesized from precursor amino acids, with dopaquinone as an intermediate metabolite. Quinones are electrophilic species and the formation of a GSH–dopaquinone conjugate serves both to "detoxify" and create an intermediate in melanin synthesis. Our own recent studies have implicated GSTP as a catalyst in the formation of the glutathionyl adduct leading to glutathionyldopa, a catalytic function which was reported earlier with GSTM2-2 (Dagnino-Subiabre et al., 2000). Sequential in the bio-synthesis, gamma–glutamyl transpeptidase and aminopeptidase N remove two of the GSH amino acids to leave cysteinyldopa, a catecholamine, excessive amounts of which have been described in melanoma bearing mice (Hu, Woodward, & Peterson, 1988) and in the plasma and urine of malignant melanoma patients(Hartleb & Arndt, 2001). Indeed, plasma cysteinyldopa has been used as a biomarker to distinguish patients with dysplastic nevi from those with primary and metastatic melanomas(Peterson et al., 1988). In addition to the pheomelanin pathway, dopaquinone can undergo slow intramolecular cyclization to cyclodopa, followed by rapid oxidation to dopachrome, the precursor of eumelanin. In the step shown, microsomal glutathione transferase 1 (MGST1) may operate as a cyclase by stabilization of the deprotonated dopaquinone (Zhang et al., 2023a).

MGST1 is a membrane bound trimeric enzyme with three GSH binding sites with generally accepted functions of membrane protection (particularly endoplasmic reticulum (ER) and outer mitochondrial (Shimoji et al., 2017)) from oxidative stress. This occurs through two catalytic activities, as a GST, and as a GSH-dependent peroxidase (GPX) (Morgenstern, Zhang, & Johansson, 2011; Zhang et al., 2023c). Evidence for linkage between MGST1 and melanogenesis is provided by several observations. For example, MGST1 is abundant in retinal pigment epithelium (Maeda, Crabb, & Palczewski, 2005), where cells have considerable numbers of melanin granules (melanosomes) and form a dark enclosure for the photoreceptor system. Melanin has antioxidant properties (Bravard, Petridis, & Luccioni, 1999; Panich et al., 2013) and its absence can result in enhanced cellular damage by ROS or other small molecule electrophilic chemicals. Transcription of MGST1 is enhanced under oxidative stress (Kelner et al., 2000), which also serves to increase antioxidant capabilities in zebra finches, leading to development of dark plumage coloration (Rodriguez-Martinez & Galvan, 2020). In lower vertebrates, melanin provides an umbrella of protection for hematopoietic stem cells, which give rise to cells of innate and adaptive immunity (Kapp et al., 2018), providing a significant link between melanin and immunity. While melanin in melanocytes serves a primary role in scavenging chemical radicals caused by UV light, melanocytes are also part of the immune system (Gasque & Jaffar-Bandjee, 2015). Despite being of distinct lineages, melanocytes share structural characteristics with dendritic cells, including branched morphology, phagocytosis, antigen processing and presentation, and production and release (Plonka et al., 2009; Tapia et al., 2014). Dendritic cells are derived from hematopoietic stem cells in the bone marrow, while melanocytes originate from neural crest cells. The localization of melanocytes in the basal stratum of the epidermis

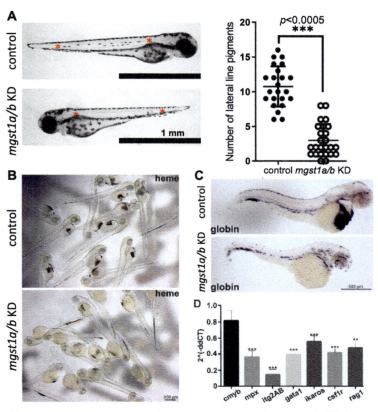

Fig. 2 Knockdown of *mgst1* expression suppresses melanin synthesis (A), hemoglobin contents (B), red blood cell numbers (C), and myeloid and lymphoid differentiation (D) in zebrafish.

provides an optimal locale for intercepting pathogens and it is likely that they work together with dendritic cells to contribute to immunity through phagocytosis (Abdallah, Mijouin, & Pichon, 2017). These characteristics illustrate roles for melanin in host immunity and provide a conduit with redox pathways that involve GST, principally GSTP and MGST1, in regulation of dendritic cell and immune functions (Brautigam et al., 2018; Zhang et al., 2014, 2018). Furthermore, in teleost zebrafish (Brautigam et al., 2018; Zhang et al., 2023a) knockdown (KD) of *mgst1* reduces the number of melanocytes localized in the midline of zebrafish embryos, and reduces myeloid and lymphoid differentiation, impacting hematopoietic stem cells and other blood cell lineages (Fig. 2), reducing immune capabilities in these fish.

3. MGST1 and melanoma

From the information provided so far, it seems reasonable to consider the importance of redox pathways, their supporting enzymes and eventual redox homeostasis in the pathogenesis of melanoma, its melanin content, the formation of metastases, and therapeutic responses. A cytosolic GST, GSTP is expressed at high levels in malignant melanoma, particularly polymorphic variants of the gene (rs1695 or 105Val; Ibarrola-Villava et al., 2012; Lei, Liu, Li, Xu, & Fan, 2015). There are indications that variants may have differential sub-cellular compartmentalization (Orlandi et al., 2009). In contrast, population-based studies suggest that altered expression of other GST isozyme families, such as GSTT and GSTM, are not associated with increased incidence of melanoma (Peng, He, Zhu, & Peng, 2013). While the literature on cytosolic GST and melanoma is quite large, information on membrane-associated GST such as MGST1 has been restricted to recent reports. Like many tumors, melanoma cells have high energy requirements necessitating high mitochondrial activity with consequent generation of ROS that are higher than their normal cell counterparts. The balance between the antioxidant properties of melanin and generated ROS creates an unusual environment where altered redox flux and homeostasis can modulate signaling events at the ER-mitochondria interface, two sub-cellular compartments where MGST1 is known to localize (Shimoji et al., 2017). High oxidative stress in the blood and visceral organs can restrict tumor spread, where establishment of metastases may depend on metabolic adaptations that increase the capacity to withstand excess oxidative stress (Piskounova et al., 2015). MGST1 levels have been shown to be a prognostic risk factor for poor survival in melanoma patients (Zeng et al., 2020b), however, specific reasons for this were not reported. Several other reports link MGST1 with cancer and are summarized in a recent review (Zhang et al., 2023c). Some examples are: increased MGST1 in ovarian carcinoma when compared to normal tissue (Hetland et al., 2012); over-expression of MGST1 linked with drug resistance, and depletion can sensitize resistant cells (Zeng et al., 2020a); MGST1 levels correlate with metastatic potential in prostate cancer (Chaib, Cockrell, Rubin, & Macoska, 2001); it is a plausible biomarker in lung cancer (Linnerth, Sirbovan, & Moorehead, 2005). The fact that MGST1 might be linked with melanin biosynthesis, oxidative stress protection, and drug response provide further rationale for considering its role in melanoma progression and metastasis, and response of melanomas to chemotherapy and immunotherapy.

By utilizing gene knockdown or over-expression experiments in highly pigmented mouse B16 or human MNT-1 melanoma cells, depletion of MGST1 was found to cause several changes in cell phenotype. These included, a marked degree of depigmentation through reduced melanin production, increased oxidative stress, slower tumor growth, and improved survival of melanoma bearing mice (Zhang et al., 2023b, 2023a). These characteristics link MGST1 with melanin biosynthesis and imply that its presence influences cell division, and/or survival. Such data are consistent with the observation that increased expression of MGST1 represents a risk gene for survival of melanoma patients. Specifically, knockdown of *MGST1* in two melanotic melanoma cell lines, B16 or MNT-1 reduced conversion of L-dopa to dopachrome (see Fig. 1), the precursor step for the indomoieties of the eumelanic polymer. There was a linear relationship between the amounts of MGST1 and the formation rates of dopachrome. Protein, mRNA and catalytic activities of tyrosinase were also significantly lower in the knockdown cells. If MGST1 was reconstituted with recombinant transfections, dramatic increases in the amount of eumelanin produced by the cells were found to occur (Zhang et al., 2023a).

4. MGST1 in melanoma metastasis and treatment response

Copious numbers of publications have detailed the relationship between acquired drug resistance and cellular alterations that involve adaptive responses in redox pathways, particularly where the expression of GSH or related thiols are concerned. For summaries of these earlier data the reader is referred to a recent comprehensive review (Hayes, Dinkova-Kostova, & Tew, 2020).

We and others have focused attention on the role of MGST1 in melanoma pathology. As a representative of the family of membrane proteins named MAPEG (membrane-associated proteins in eicosanoid and glutathione metabolism), it is found extensively in ER and outer mitochondrial membranes. Enzyme substrates are usually hydrophobic and MGST1 plays a role in the biotransformation of lipophilic reactive electrophiles and reduction of membrane-embedded phospholipid hydroperoxides, displaying both GST and GPX activities (Morgenstern et al., 2011). Expression at two distinct intracellular locations is an unusual feature for a membrane protein, suggesting a role for MGST1 in protecting ER and outer mitochondrial membranes from lipid peroxidation and a downstream protective impact on prevention of ferroptosis. In this regard, MGST1 has been shown to bind to ALOX5, suppressing ferroptosis by reducing the generation of lipid peroxides (Kuang, Liu, Xie, Tang, & Kang, 2021). While enhanced drug response and decreased metastasis through inhibition of GPX4 has been suggested as a strategy for melanoma therapy, the role of MGST1 in melanoma pathology has only recently been considered.

In our recent study, we identified MGST1 as highly expressed in dedifferentiated and drug resistant human melanomas and as a specific determinant of metastatic spread and therapeutic sensitivity (Zhang et al., 2023b). Inhibition of MGST1 results in increased lipid peroxidation and enhanced levels of ferroptosis. Knockdown of *Mgst1* was shown to enhance the efficacy of both chemotherapy and immunotherapy in melanotic melanomas. For example, co-culture of melanoma cells with gp100 acti-vated splenic Pmel T cells, resulted in greater cytotoxicity in *Mgst1* KD cells and elevated expression of chemokine levels (CCL5, CCL6, CCL8, CCL9/10, CCL21, CXCL1, Comp FD, C5a, chemerin, and IL16). When *Mgst1* KD was induced in these melanoma cells, increased migration of both splenic T cells from C57BL/6 mice and gp100-activated Pmel T cells occurred. For gp100-activated Pmel T cells, increased chemokine-induced trafficking of $CD3^+$ and $CD8^+$ T cells towards melanoma cells was pro-nounced. In addition, *Mgst1* KD B16 cells were more sensitive to a range of anticancer drugs with quite distinct and unrelated mechanisms of action. These included: vinblastine (tubulin destabilization), paclitaxel (tubulin stabilization) (Shah & Dronca, 2014), doxorubicin (topoisomerase inhi-bitor) (Zhang et al., 2015), PABA/NO (GSTP activated prodrug releasing nitric oxide) (Saavedra et al., 2006) or ME344 (mitochondria targeted isoflavone) (Zhang et al., 2019a, 2019b) (Fig. 3). A common feature of these drugs is their capacity to produce (either directly or indirectly) ROS or RNS (Hayes et al., 2020), consistent with the interpretation that decreased melanin concentrations in *MGST1* KD cells diminished their capacity to scavenge ROS, with resultant increased drug toxicities. Genetic down-regulation of *Mgst1* sensitizes melanoma to drug and immune therapies. If small molecule inhibitors of MGST1 could be identified, these could be of utility with respect to melanoma treatment response.

The expression of melanin has been reported to affect the elastic properties of the cells as well as their migration capabilities (Slominski et al., 2015). Thus, unsurprisingly, *MGST1* KD was found to suppress both melanoma cell migration *in vitro* and establishment of

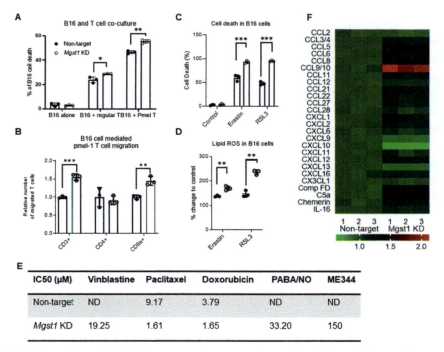

Fig. 3 Knockdown (KD) of *Mgst1* was shown to enhance the efficacy of chemotherapy and immunotherapy in B16 melanotic melanomas. (A) and (B) Co-culture of melanoma cells with regular and gp100 activated Pmel T cells. (C–E) Enhanced cellular sensitivity to cytotoxic anticancer drugs and ferroptotic cell death. *ND*, not detectable. (F) Elevated expression of chemokine levels in *Mgst1* KD B16 cells.

metastatic lesions in C57BL/6 mice. Morphological changes were observed in both B16 and MNT-1 *MGST1* KD cells where KD cells were smaller and more condensed, with fewer signs of dendrite formation. Melanoma cells commonly metastasize to the lung, generating surface lesions detectable as pigmented spots. *Mgst1* KD interfered with the formation of lung metastases, where higher recruitment of $CD8^+$ T and dendritic cells ($CD11c^+$) to the tumor sites also occurred. The expression of CXCR3, which correlates with the improved migration of the anti-tumor T cells to the tumor site, was elevated in the T cells from spleen of B16 *Mgst1* KD tumor-bearing mice. Mice injected with B16 cells died between 19- and 22-days following inoculation. However, mice inoculated with B16 *Mgst1* KD cells had enhanced survival, with only 10% dead at 26 days, and 30% surviving through a 42-day observation period. Overall, these results amplify the importance of the

Redox pathways in melanoma 135

multiple functions of melanin and connect them with melanoma progression. MGST1 inhibitors may provide a potential therapeutic approach in the management of metastatic melanoma.

5. Additional redox targets in melanoma

In addition to those discussed above, several redox associated pathways are dysregulated in melanoma and a variety of reports implicates them in melanoma occurrence, treatment, and resistance. These are summarized in this section.

As a bioavailable version of cysteine, N-acetylcysteine (NAC) pretreatment was found to block formation of 8-oxoguanine in mouse skin following neonatal UV treatment and delayed onset of UV-induced melanotic tumors (Goodson et al., 2009). NAC increased the thiol pools, attenuated UV radiation induced GSH depletion and diminished UV-induced oxidative stress in nevi. However, NAC has also been shown to increase human melanoma cell migration and division, accelerating metastasis in murine models of malignant melanoma. Perversely, high doses of NAC can increase the generation of ROS and subsequently the nuclear translocation of NRF2. This is likely the consequence of induction of reductive stress (Zhang et al., 2024), a process that overloads the reductive potential of cells and can lead to imbalances in NADH and GSH (Zhang & Tew, 2021). As with many other examples of the biological uses for NAC, the dose effects and paradoxical nature of how NAC influences oxidative stress pathways should be viewed with caution.

In terms of redox homeostasis, cysteine and GSH act as major sources of reduced thiols in cells and their active transport, generally by members of the "SLC" family of transporters is necessary for maintenance of organelles such as the melanosome (Potterf et al., 1999) and mitochondria(Liu et al., 2023), where SLC7A11 has been identified as a primary transporter across membranes. Its expression is enhanced in many cancer types (including melanoma), and this can serve to alter intracellular redox balance and mitigate the cytotoxic effects of elevated oxidative stress associated with this oncogenic phenotype. SLC7A11 expression can be elevated in melanoma, both before and after the acquisition of drug resistance (Shin et al., 2018). The cysteine cycle has recently been identified as a potential target for melanoma therapy, as inhibiting the function of xCT (cysteine-glutamate antiporter; system X_c^-, comprised of two subunits, SLC7A11/xCT and

SLC3A2) with (*S*)-4-carboxyphenylglycine increased cytotoxicity in melanoma cells, especially when used in combination with inhibition of GSH synthesis using buthionine sulfoximine (Vene et al., 2011). The xCT antiporter has been shown to contribute both *in vitro* and *in vivo* towards melanoma proliferation, and a small molecule inhibitor, riluzole interferes with proliferation (Shin et al., 2018). Moreover, knockout of SLC7A11 using CRISPR/Cas9 technology inhibits proliferation, migration, and invasion *in vitro*, and in mouse xenografts interference with cysteine transport also facilitated proliferation and metastasis (Chen et al., 2023). Expression of SLC7A11 is also increased in melanoma cell lines made resistant to BRAF inhibitors (BRAFi), which themselves exhibit elevated levels of oxidative stress as compared to drug naive cells (Wang et al., 2018). In melanoma patients, treatment with MAPK inhibitors and vorinostat, a histone deacetylase inhibitor, suppressed SLC7A11 expression in primary and metastatic lesions. Generally, such results signify a potential for targeting SLC7A11, perhaps in combination with other therapies to counteract the altered redox homeostasis present in melanomas both before and after acquisition of drug resistance phenotype(s).

As in most tumors, rewiring of metabolic pathways can help to sustain proliferation and antioxidant defenses, but also may provide therapeutic opportunities (Martinez-Outschoorn, Peiris-Pages, Pestell, Sotgia, & Lisanti, 2017; Wolpaw & Dang, 2018). Melanoma cells can reprogram amino acid and lipid metabolism and use alternate energy production pathways (Fischer et al., 2018). BRAF mutations occur in approximately 60% of melanoma patients and activation of the BRAF/MAPK pathways negatively regulates oxidative metabolism *via* PGC1α and MITF. Treatment with BRAF BRAFi can convert melanoma cells to be addicted to oxidative phosphorylation, limiting the efficacy of BRAFi (Schockel et al., 2015). BRAFi-resistant melanomas are more sensitive to cell death induced by mitochondrial-targeting drugs. Inhibition of mitochondrial complex I reduces melanoma tumor growth by inducing mPTP opening and mitochondrial membrane potential and depolarization, stimulating autophagosome formation, mitophagy and associated ROS increases, leading to activation of combined necroptotic/ferroptotic cell death (Basit et al., 2017). In addition, BRAFi-resistance can alter mitochondrial metabolism to support glucose-derived glutamate synthesis. BRAFi-resistant melanoma efficiently activates Nrf2, leading to increased pentose phosphate pathway (PPP), which can regenerate GSH, and increase xCT expression, stimulating uptake of cystine required for intracellular GSH synthesis. These metabolic adaptations can sustain intracellular GSH and redox homeostasis to facilitate BRAFi-resistant melanoma survival. In part consequently, BRAFi-resistant melanoma is more sensitive to drugs targeting GSH metabolism and ferroptosis (Khamari et al., 2018). In addition, sterol regulatory binding element 1 (SREBP-1) mediated induction of lipid biosynthesis promotes lipid saturation, protecting melanoma from lipid peroxidation, and contributing to therapy resistance in BRAF mutant melanoma. SREBP-1 inhibition sensitizes melanoma cells to BRAFi, partly through inducing ferroptosis (Talebi et al., 2018). Metabolic adaptations within a population of cells can contribute to differences in metastatic potential, which can be delimited by increased oxidative stress. Melanoma cells that are successful in establishing metastasis undergo reversible metabolic changes during the metastasis process that increase their capacity to deal with oxidative stress, including increased dependence on both the NADPH-generating enzymes of the folate pathway (Piskounova et al., 2015) and the monocarboxylate transporter 1 in circulating lactate for PPP (Tasdogan et al., 2020). Ferroptosis evasion through the modulation of fatty acid metabolism can also contribute to cancer metastasis. The lymphatic environment with higher levels of GSH and oleic acid and reduced levels of free iron can protect melanoma cells from ferroptosis and increase their survival during subsequent metastasis into the bloodstream. Oleic acid protects melanoma cells from ferroptosis contingent upon acyl-CoA Synthetase Long Chain Family Member 3 (ACSL3). ACSL3-mediated monounsaturated fatty acid synthesis suppresses ferroptosis by displacing polyunsaturated fatty acids from phospholipids (Ubellacker et al., 2020).

In common with several cancers, the expression of both thioredoxin (TRX) and thioredoxin reductase (TR) can be elevated in melanoma (Carpenter et al., 2022). TRX levels seem to increase commensurate with disease progression, and elevated expression has been linked, perhaps functionally, with heightened incidence of tumor metastasis. TRX secretion from melanoma cells is a contributory factor in cancer immune evasion, where recruitment and differentiation of T cells into immunosuppressive regulatory T cells is inhibited (Soderberg, Sahaf, & Rosen, 2000). A negative regulator of TRX, thioredoxin interacting protein (TXNIP), is present at substantially lower levels in a graduation from benign nevi to primary melanoma and are even further reduced in metastatic melanoma (Cheng et al., 2004). In this same report, TXNIP expression was elevated following treatment with a BRAF inhibitor, PLX4032, and then decreased again when cells acquired resistance to the drug. These types of data are indicative of a cause: effect relationship between melanoma progression and small molecule redox regulators exemplified by the TRX family. The expression of TR1 was analyzed in tissue microarrays of human nevi and melanomas and was found to increase as a function of disease progression (Cassidy et al., 2015). Melanoma cell lines had metabolic differences that correlated with TR1 levels. This study used synthetic lethal targeting of TR1 which they found sensitized melanoma to inhibition of glycolytic metabolism, causing a decrease in metastases in mice. As an interesting corollary, enhanced pigmentation in guinea pigs was found to be accompanied by increased expression of TR1, at the same time as other antioxidant enzymes, superoxide dismutase (cytosol Cu/Zn-enzyme), catalase, and glutathione reductase, decreased. When eumelanin biosynthesis was induced in murine melanoma cells by isobutylmethylxanthine, TR1 levels also increased coincident with melanogenesis. The authors suggested that both tyrosinase and TR1 respond to oxidative stress in the epidermis and/ or melanoma cells and react with superoxide anion radicals to stimulate melanogenesis and to prevent peroxidative damage, respectively (Schallreuter, Lemke, Hill, & Wood, 1994).

In uveal melanoma (UM), the presence of cancers over-expressing mitochondrial thioredoxin-dependent peroxide reductase (PRDX3) has been an indicator of poor prognosis, with high expression correlated to metastasis and reduced patient survival (Ramasamy et al., 2020). Although occurrences are low in the general population, UM is the most common primary intraocular malignancy in adults, and 40% develop fatal metastatic disease. Over-expression of PRDX3 has been described in several types of cancers including, prostate, breast, colorectal and lung cancer. In a study of 92 uveal cancer patients, high immunohistochemical expression of PRDX3 in primary tumor tissue was associated with metastasis and poor survival (Ramasamy et al., 2020).

Further correlative analyses in a UM patient cohort revealed that increased median levels of heme oxygenase 1 (HO-1) expression were found in metastasized UM when compared to patients without metastasis. The authors suggested that HO-1 derived carbon monoxide might be involved in UM progression and HO-1 protein expression may be of potential prognostic and therapeutic value in these patients (Longhitano et al., 2022). In addition, preclinical approaches determined that HO-1 was associated with a marked activation of BRAF-ERK1/2 signaling and led to CDK2/cyclin E activation, thereby promoting melanoma proliferation, implying that the HO-1/BRAF/ERK axis may play a role in melanoma cell proliferation and that targeting HO-1 could be a therapeutic approach for treating melanoma (Liu et al., 2019).

Several unrelated articles describe other redox components that may have an influence of the development or treatment of melanoma. These are summarized in the following section.

Superoxide dismutases (SOD) have been considered as promising targets in melanomas or overcoming drug resistance by interfering with antioxidant protection of the mitochondria (Benlloch et al., 2006). Inhibition of CuZnSOD (SOD1) by disulfiram improved the chemosensitivity of melanoma cells to subsequent oxaliplatin treatment. A combination of curcumin and disulfiram (or some of its metabolites) displayed synergistic tumor growth inhibition (Rieber, 2020). Also, curcumin is known to stimulate ROS production, reported to induce caspase activation leading to more cell death and apoptosis in melanoma cells when compared to normal cells (Manica et al., 2023). Given the complexity of off-target effects for both disulfiram and curcumin, cautionary interpretation of any redox specific effects from such studies are recommended.

Given the relative importance of selenocysteine in the structure/function of a small group of critical antioxidant enzymes, several population-based studies have documented the relevance of selenium as an indicator of melanoma occurrence, progression, or response. One of the recent reports concludes that low serum selenium levels were associated with increased mortality rates in the 10 years following melanoma diagnosis (Rogoza-Janiszewska et al., 2021). Once again, given the non-specific effects of selenium, specific conclusions are more difficult to attain. There is controversy as to whether selenium supplementation may have value in melanoma prevention and/or treatment. Topical treatment with L-seleno-methionine caused a delay in the time required for UV-induced melanoma development, but also increased the rate of growth of those tumors once they appeared. There was no indication that the incidence of cutaneous melanoma was impacted (Cassidy et al., 2013).

As mentioned earlier, reducing ROS with NAC promoted melanoma cell migration and metastasis. A recent report considered the generality of this antioxidant effect by screening 104 redox-active compounds and identified 18 that increased migration of human malignant melanoma cells. Amongst these, vitamin C ranked as number one, followed by the vitamin E analogue Trolox and several carotenoids and Vitamin A-related compounds. Four diet-relevant compounds from this list, including vitamin C were found to accelerate the rate of formation of distant metastases in mice with BRAF(V600E)-driven malignant melanoma. The study identified the transcription factor BACH1 as activated following antioxidants and knockout of BACH1 in mouse melanoma cells reduced lymph node and liver metastases. Thus, it seems that a broad range of antioxidants can influence the rate of melanoma migration and metastasis and that BACH1 may mediate such a function (Kashif et al., 2023). As a general conclusion in humans, vitamin supplementation, particularly those that fall into the antioxidant category, should be placed in perspective with respect to the doses administered.

Acknowledgments

This work was supported by grants from National Institutes of Health 5P20GM103542 — South Carolina COBRE in Oxidants, Redox Balance and Stress Signaling, Department of Defense office of the Congressionally Directed Medical Research Programs through the Melanoma Research Program under Award Number HT9425-23-1-0649, ACURO Protocol Number ME220116.e001, the South Carolina Centers of Economic Excellence program, and American Cancer Society Institutional Research Grant (ACS-IRG) IRG-19-137-20.

References

Abdallah, F., Mijouin, L., & Pichon, C. (2017). *Mediators of Inflammation, 2017*, 5095293.
Basit, F., van Oppen, L. M., Schockel, L., Bossenbroek, H. M., van Emst-de Vries, S. E., Hermeling, J. C., ... Koopman, W. J. (2017). *Cell Death & Disease, 8*, e2716.
Becker, A. L., & Indra, A. K. (2023). *Cancers (Basel), 15*.
Benlloch, M., Mena, S., Ferrer, P., Obrador, E., Asensi, M., Pellicer, J. A., ... Estrela, J. M. (2006). *The Journal of Biological Chemistry, 281*, 69–79.
Brautigam, L., Zhang, J., Dreij, K., Spahiu, L., Holmgren, A., Abe, H., ... Johansson, K. (2018). *Redox Biology, 17*, 171–179.
Bravard, A., Petridis, F., & Luccioni, C. (1999). *Free Radical Biology & Medicine, 26*, 1027–1033.
Brozyna, A. A., Jozwicki, W., Carlson, J. A., & Slominski, A. T. (2013). *Human Pathology, 44*, 2071–2074.
Carpenter, E. L., Becker, A. L., & Indra, A. K. (2022). *Cancers (Basel), 14*.
Cassidy, P. B., Fain, H. D., Cassidy, J. P., Jr., Tran, S. M., Moos, P. J., Boucher, K. M., ... Leachman, S. A. (2013). *Nutrients, 5*, 725–749.
Cassidy, P. B., Honeggar, M., Poerschke, R. L., White, K., Florell, S. R., Andtbacka, R. H., ... Moos, P. J. (2015). *Pigment Cell & Melanoma Research, 28*, 685–695.
Chaib, H., Cockrell, E. K., Rubin, M. A., & Macoska, J. A. (2001). *Neoplasia (New York, N. Y.), 3*, 43–52.
Cheng, G. C., Schulze, P. C., Lee, R. T., Sylvan, J., Zetter, B. R., & Huang, H. (2004). *Experimental Cell Research, 300*, 297–307.
Chen, Y., Lu, T., Liu, Y., Liu, Y., Bai, S., Chen, Q., ... Wu, X. (2023). *In Vitro Cellular & Developmental Biology. Animal, 59*, 729–737.
Dagnino-Subiabre, A., Cassels, B. K., Baez, S., Johansson, A. S., Mannervik, B., & Segura-Aguilar, J. (2000). *Biochemical and Biophysical Research Communications, 274*, 32–36.
Fischer, G. M., Vashisht Gopal, Y. N., McQuade, J. L., Peng, W., DeBerardinis, R. J., & Davies, M. A. (2018). *Pigment Cell & Melanoma Research, 31*, 11–30.

Gasque, P., & Jaffar-Bandjee, M. C. (2015). *The Journal of Infection, 71*, 413–421.

Goodson, A. G., Cotter, M. A., Cassidy, P., Wade, M., Florell, S. R., Liu, T., ... Grossman, D. (2009). *Clinical Cancer Research: An Official Journal of the American Association for Cancer Research, 15*, 7434–7440.

Hartleb, J., & Arndt, R. (2001). *Journal of Chromatography. B, Biomedical Sciences and Applications, 764*, 409–443.

Hayes, J. D., Dinkova-Kostova, A. T., & Tew, K. D. (2020). *Cancer Cell, 38*, 167–197.

Hetland, T. E., Nymoen, D. A., Emilsen, E., Kaern, J., Trope, C. G., Florenes, V. A., & Davidson, B. (2012). *Gynecologic Oncology, 126*, 460–465.

Hu, F., Woodward, W. R., & Peterson, L. L. (1988). *The Journal of Investigative Dermatology, 90*, 149–151.

Ibarrola-Villava, M., Martin-Gonzalez, M., Lazaro, P., Pizarro, A., Lluch, A., & Ribas, G. (2012). *The British Journal of Dermatology, 166*, 1176–1183.

Kapp, F. G., Perlin, J. R., Hagedorn, E. J., Gansner, J. M., Schwarz, D. E., O'Connell, L. A., ... Zon, L. I. (2018). *Nature, 558*, 445–448.

Kashif, M., Yao, H., Schmidt, S., Chen, X., Truong, M., Tuksammel, E., ... Bergo, M. O. (2023). *Redox Biology, 60*, 102619.

Kelner, M. J., Bagnell, R. D., Montoya, M. A., Estes, L. A., Forsberg, L., & Morgenstern, R. (2000). *The Journal of Biological Chemistry, 275*, 13000–13006.

Khamari, R., Trinh, A., Gabert, P. E., Corazao-Rozas, P., Riveros-Cruz, S., Balayssac, S., ... Kluza, J. (2018). *Cell Death & Disease, 9*, 325.

Kuang, F., Liu, J., Xie, Y., Tang, D., & Kang, R. (2021). *Cell Chemical Biology, 28*, 765–775 e5.

Lei, Z., Liu, T., Li, X., Xu, X., & Fan, D. (2015). *International Journal of Clinical and Experimental Medicine, 8*, 377–386.

Linnerth, N. M., Sirbovan, K., & Moorehead, R. A. (2005). *International Journal of Cancer, 114*, 977–982.

Liu, Y., Liu, S., Tomar, A., Yen, F. S., Unlu, G., Ropek, N., ... Birsoy, K. (2023). *Science (New York, N. Y.), 382*, 820–828.

Liu, L., Wu, Y., Bian, C., Nisar, M. F., Wang, M., Hu, X., ... Zhong, J. L. (2019). *Cell Communication and Signaling: CCS, 17*, 3.

Longhitano, L., Broggi, G., Giallongo, S., Failla, M., Puzzo, L., Avitabile, T., ... Musso, N. (2022). *Antioxidants (Basel), 11*.

Maeda, A., Crabb, J. W., & Palczewski, K. (2005). *Biochemistry, 44*, 480–489.

Manica, D., Silva, G. B. D., Silva, A. P. D., Marafon, F., Maciel, S., Bagatini, M. D., & Moreno, M. (2023). *Cell Biochemistry and Function, 41*, 1295–1304.

Martinez-Outschoorn, U. E., Peiris-Pages, M., Pestell, R. G., Sotgia, F., & Lisanti, M. P. (2017). *Nature Reviews Clinical Oncology, 14*, 11–31.

Morgenstern, R., Zhang, J., & Johansson, K. (2011). *Drug Metabolism Reviews, 43*, 300–306.

Orlandi, A., Costantini, S., Campione, E., Ferlosio, A., Amantea, A., Bianchi, L., ... Spagnoli, L. G. (2009). *Archives of Dermatology, 145*, 55–62.

Panich, U., Pluemsamran, T., Tangsupa-a-nan, V., Wattanarangsan, J., Phadungrakwittaya, R., Akarasereenont, P., & Laohapand, T. (2013). *BMC Complementary and Alternative Medicine, 13*, 159.

Peng, H., He, Q., Zhu, J., & Peng, C. (2013). *Tumour Biology: The Journal of the International Society for Oncodevelopmental Biology and Medicine, 34*, 675–681.

Peterson, L. L., Woodward, W. R., Fletcher, W. S., Palmquist, M., Tucker, M. A., & Ilias, A. (1988). *Journal of the American Academy of Dermatology, 19*, 509–515.

Piskounova, E., Agathocleous, M., Murphy, M. M., Hu, Z., Huddlestun, S. E., Zhao, Z., ... Morrison, S. J. (2015). *Nature, 527*, 186–191.

Plonka, P. M., Passeron, T., Brenner, M., Tobin, D. J., Shibahara, S., Thomas, A., ... Schallreuter, K. U. (2009). *Experimental Dermatology, 18*, 799–819.

Potterf, S. B., Virador, V., Wakamatsu, K., Furumura, M., Santis, C., Ito, S., & Hearing, V. J. (1999). *Pigment Cell Research / Sponsored by the European Society for Pigment Cell Research and the International Pigment Cell Society, 12*, 4–12.

Ramasamy, P., Larkin, A. M., Linge, A., Tiernan, D., McAree, F., Horgan, N., ... Meleady, P. (2020). *Journal of Clinical Pathology, 73*, 408–412.

Rieber, M. (2020). *Current Pharmaceutical Design, 26*, 4461–4466.

Rodriguez-Martinez, S., & Galvan, I. (2020). *Comparative Biochemistry and Physiology. Toxicology & Pharmacology: CBP, 228*, 108667.

Rogoza-Janiszewska, E., Malinska, K., Baszuk, P., Marciniak, W., Derkacz, R., Lener, M., ... Lubinski, J. (2021). *Biomedicines, 9*.

Saavedra, J. E., Srinivasan, A., Buzard, G. S., Davies, K. M., Waterhouse, D. J., Inami, K., ... Keefer, L. K. (2006). *Journal of Medicinal Chemistry, 49*, 1157–1164.

Schallreuter, K. U., Lemke, K. R., Hill, H. Z., & Wood, J. M. (1994). *The Journal of Investigative Dermatology, 103*, 820–824.

Schockel, L., Glasauer, A., Basit, F., Bitschar, K., Truong, H., Erdmann, G., ... Heroult, M. (2015). *Cancer Metabolism, 3*, 11.

Shah, D. J., & Dronca, R. S. (2014). *Mayo Clinic Proceedings. Mayo Clinic, 89*, 504–519.

Shimoji, M., Figueroa, R. A., Neve, E., Maksel, D., Imreh, G., Morgenstern, R., & Hallberg, E. (2017). *Biochimica et Biophysica Acta (BBA) - Biomembranes, 1859*, 238–244.

Shin, S. S., Jeong, B. S., Wall, B. A., Li, J., Shan, N. L., Wen, Y., ... Chen, S. (2018). *Oncogenesis, 7*, 86.

Slominski, A. T., & Carlson, J. A. (2014). *Mayo Clinic Proceedings. Mayo Clinic, 89*, 429–433.

Slominski, R. M., Zmijewski, M. A., & Slominski, A. T. (2015). *Experimental Dermatology, 24*, 258–259.

Soderberg, A., Sahaf, B., & Rosen, A. (2000). *Cancer Research, 60*, 2281–2289.

Talebi, A., Dehairs, J., Rambow, F., Rogiers, A., Nittner, D., Derua, R., ... Swinnen, J. V. (2018). *Nature Communications, 9*, 2500.

Tapia, C. V., Falconer, M., Tempio, F., Falcon, F., Lopez, M., Fuentes, M., ... Di Nardo, A. (2014). *Medical Mycology: Official Publication of the International Society for Human and Animal Mycology, 52*, 445–454.

Tasdogan, A., Faubert, B., Ramesh, V., Ubellacker, J. M., Shen, B., Solmonson, A., ... Morrison, S. J. (2020). *Nature, 577*, 115–120.

Ubellacker, J. M., Tasdogan, A., Ramesh, V., Shen, B., Mitchell, E. C., Martin-Sandoval, M. S., ... Morrison, S. J. (2020). *Nature, 585*, 113–118.

Vene, R., Castellani, P., Delfino, L., Lucibello, M., Ciriolo, M. R., & Rubartelli, A. (2011). *Antioxidants & Redox Signaling, 15*, 2439–2453.

Wang, L., Leite de Oliveira, R., Huijberts, S., Bosdriesz, E., Pencheva, N., Brunen, D., ... Bernards, R. (2018). *Cell, 173*, 1413–1425 e14.

Wolpaw, A. J., & Dang, C. V. (2018). *Trends in Cell Biology, 28*, 201–212.

Zeng, B., Ge, C., Li, R., Zhang, Z., Fu, Q., Li, Z., ... Huang, Y. (2020a). *Biomedicine & Pharmacotherapy, 121*, 109562.

Zeng, F., Su, J., Peng, C., Liao, M., Zhao, S., Guo, Y., ... Deng, G. (2020b). *Frontiers in Oncology, 10*, 1710.

Zhang, X., Teodoro, J. G., & Nadeau, J. L. (2015). *Nanomedicine: Nanotechnology, Biology, and Medicine, 11*, 1365–1375.

Zhang, L., & Tew, K. D. (2021). *Advances in Cancer Research, 152*, 383–413.

Zhang, J., Ye, Z. W., Brautigam, L., Chakraborty, P., Luo, Z., Culpepper, J., ... Tew, K. D. (2023a). *The Journal of Biological Chemistry, 299*, 104920.

Zhang, J., Ye, Z. W., Chakraborty, P., Luo, Z., Culpepper, J., Aslam, M., ... Tew, K. D. (2023b). *Pharmacological Research: The Official Journal of the Italian Pharmacological Society, 196*, 106899.

Zhang, J., Ye, Z. W., Chen, W., Manevich, Y., Mehrotra, S., Ball, L., ... Townsend, D. M. (2018). *The Journal of Biological Chemistry, 293*, 4366–4380.

Zhang, J., Ye, Z. W., Gao, P., Reyes, L., Jones, E. E., Branham-O'Connor, M., ... Tew, K. D. (2014). *PLoS One, 9*, e107478.

Zhang, J., Ye, Z. W., Morgenstern, R., Townsend, D. M., & Tew, K. D. (2023c). *Advances in Cancer Research, 160*, 107–132.

Zhang, L., Zhang, J., Ye, Z., Manevich, Y., Ball, L. E., Bethard, J. R., ... Tew, K. D. (2019a). *Cancer Research, 79*, 4072–4085.

Zhang, L., Zhang, J., Ye, Z. W., Muhammad, A., Li, L., Culpepper, J. W., ... Tew, K. D. (2024). *Biochemical Pharmacology, 219*, 115929.

Zhang, L., Zhang, J., Ye, Z., Townsend, D. M., & Tew, K. D. (2019b). *Advances in Cancer Research, 142*, 187–207.

CHAPTER SIX

Melanoma redox biology and the emergence of drug resistance

Therese Featherston, Martina Paumann-Page*, and Mark B. Hampton*

Mātai Hāora—Centre for Redox Biology and Medicine, Department of Pathology and Biomedical Science, University of Otago, Christchurch, New Zealand
*Corresponding authors. e-mail address: martina.paumann-page@otago.ac.nz; mark.hampton@otago.ac.nz

Contents

1. Biology of melanoma	147
2. Treatment of melanoma	148
3. Drug resistance in melanoma	149
4. Oxidative stress in melanoma	152
5. Antioxidants in melanoma	155
6. Oxidative stress and BRAF inhibitor resistance	159
Acknowledgments	163
References	163

Abstract

Melanoma is the deadliest form of skin cancer, with the loss of approximately 60,000 lives world-wide each year. Despite the development of targeted therapeutics, including compounds that have selectivity for mutant oncoproteins expressed only in cancer cells, many patients are either unresponsive to initial therapy or their tumors acquire resistance. This results in five-year survival rates of below 25%. New strategies that either kill drug-resistant melanoma cells or prevent their emergence would be extremely valuable. Melanoma, like other cancers, has long been described as being under increased oxidative stress, resulting in an increased reliance on antioxidant defense systems. Changes in redox homeostasis are most apparent during metastasis and during the metabolic reprogramming associated with the development of treatment resistance. This review discusses oxidative stress in melanoma, with a particular focus on targeting antioxidant pathways to limit the emergence of drug resistant cells.

Abbreviations

ABC	ATP-binding cassette
Akt	protein kinase B
AMP	adenosine monophosphate
AMPK	AMP-activated protein kinase
ARE	antioxidant response element
ASK1	apoptosis signal-regulating kinase 1

ATF4	activating transcription factor 4
CHORDC1	cysteine- and histidine-rich domain-containing protein 1
DUOX	dual oxidase
EMT	epithelial-mesenchymal transition
ERK	extracellular signal-regulated kinase
FOXO	forkhead box-O
GLRX	glutaredoxin
GPX	glutathione peroxidase
GR	glutathione reductase
GSH	glutathione
GSSG	oxidized glutathione disulfide
GTP	guanosine triphosphate
H$_2$O$_2$	hydrogen peroxide
HOBr	hypobromous acid
IGF1R	insulin-like growth factor 1 receptor
MC1R	melanocortin 1 receptor
MEK	mitogen activated protein kinase kinase
MITF	microphthalmia-associated transcription factor
mTOR	mammalian target of rapamycin
αMSH	alpha-melanocyte-stimulating hormone
NADPH	nicotinamide adenine dinucleotide phosphate
NF-κB	nuclear factor kappa-light-chain-enhancer of activated B cells
NFKB2	NFκB p100 subunit 2
NOX	NADPH oxidase
NRF2	nuclear factor erythroid 2-related factor 2
PDGFR-β	platelet derived growth factor receptor beta
PDH	pyruvate dehydrogenase
PDK	pyruvate dehydrogenase kinase
PGC1α	peroxisome proliferator-activated receptor-gamma coactivator 1 alpha
PI3K	phosphoinositide 3-kinase
PRDX	peroxiredoxin
PXDN	peroxidasin
Rb	retinoblastoma
SNAI1	snail family transcriptional repressor 1
SOD	superoxide dismutase
TCA	tricarboxylic acid
TRi-1	thioredoxin reductase 1 inhibitor
TRi-2	thioredoxin reductase 2 inhibitor
TXN	thioredoxin
TXNIP	thioredoxin interacting protein
TXNRD	thioredoxin reductase
UV	ultraviolet

1. Biology of melanoma

Melanomas originate from pigment-producing melanocytes localized in the basal layer of the skin epidermis (Damsky et al., 2010). Approximately 30% arise from melanocytic nevus precursors, more commonly known as moles, while the remainder arise spontaneously (Lin et al., 2015). Surgical excision of localized melanoma is essentially curative, but advanced metastatic melanoma is notoriously aggressive and resistant to treatment (Smithers et al., 2021). Regional metastatic sites of melanoma are typically nearby skin and lymph nodes, while lung, brain and liver are the common distant metastatic sites (Damsky et al., 2010).

Melanoma is one of the most genetically heterogeneous cancers with multiple mutations identified across different cases (Tirosh et al., 2016; Tran et al., 2021). This high mutation load is largely due to ultraviolet (UV) radiation–induced mutations, although the two most common driver mutations in melanoma, affecting the *BRAF* and *NRAS* genes, are not UV-induced (Hodis et al., 2012). The *BRAF* oncogene encodes for the B-Raf serine/threonine-protein kinase, which forms part of the Ras-Raf-mitogen activated protein kinase kinase (MEK)-extracellular signal–regulated kinase (ERK) signaling pathway that controls cell cycle progression. Mutations in the *BRAF* gene are found in approximately 50% of melanomas (Davies et al., 2002), with the most common variant being V600E, caused by a T1779A point mutation. This mutation results in the protein always being in its conformational active state (Bollag et al., 2012), leading to continuous phosphorylation of downstream MEK and ERK kinases, hyperphosphorylation of retinoblastoma (Rb) protein and nuclear translocation of E2F transcription factor (Yang, Tian, Hoffman, Jacobsen, & Spencer, 2021). E2F promotes transcription of downstream genes that regulate G1 to S phase of the cell cycle, resulting in uncontrolled cell division, differentiation and proliferation.

A mutation in the *NRAS* gene is found in approximately 20% of melanoma patients (Hodis et al., 2012). The *NRAS* gene encodes for the N-Ras GTPase, which is also involved the Ras-Raf-MEK-ERK signaling pathway, and the phosphoinositide 3-kinase (PI3K)/protein kinase B (Akt)/mammalian target of rapamycin (mTOR) signaling pathway that controls cell cycle progression (Flaherty, Hodi, & Fisher, 2012). The most prevalent mutation is Q61R/K (Akbani et al., 2015), which leads to constitutive activation of the Ras-Raf-MEK-ERK signaling pathway.

2. Treatment of melanoma

Current treatment for melanoma is based on the stage of disease. If detected early, while clear boundaries of the primary tumor remain and no metastases are identified, then surgery to excise the primary tumor is performed (Smithers et al., 2021). This can be curative, particularly if a large margin is achieved around the primary tumor boundary. Sometimes, adjuvant therapies such as radiotherapy or chemotherapy are used to reduce the risk of recurrence. As patients present with more advanced stages of disease, where the primary tumor has begun to invade the basement membrane and metastasis has occurred, then treatment options are more limited. Frequent and often undetected melanoma metastasis along with the limited treatment options make melanoma the most fatal form of skin cancer.

The development of targeted therapeutics and immunotherapy have proved successful for melanoma patients (Flaherty et al., 2012). Targeted therapy describes a form of treatment where inhibitors specific to cancer-inducing mutant proteins are used to effectively inhibit cell proliferation and trigger cell death. For melanoma, drugs have been developed that specifically bind to the BRAF V600E oncoprotein. The first to be developed was vemurafenib, followed by dabrafenib and more recently encorafenib (Fig. 1) (Bollag et al., 2012). Vemurafenib selectively binds to the cleft between the N- and C-terminal lobes of the kinase domain that overlaps the ATP-binding site (Tsai et al., 2008). This results in outward movement of the regulatory αC helix that flanks the ATP-binding site, interfering with the dimerization of the active-state protein and inhibiting kinase activity (Bollag et al., 2012). Vemurafenib binds most strongly when BRAF is in its active conformation, meaning that it has limited impact on wildtype BRAF protein (Tsai et al., 2008). Initial results indicated great success for late stage melanoma patients with this specific mutation, with survival ranging from six to 10 months and response rates around 50% (Chapman et al., 2011). Dabrafenib (Rheault et al., 2013) and encorafenib (Koelblinger, Thuerigen, & Dummer, 2018; Schadendorf et al., 2024) were subsequently identified as more effective BRAF inhibitors, and have also been approved for late stage melanoma treatment.

MEK can be constitutively phosphorylated by BRAF mutant-independent mechanisms, for example mutations of Ras (Corcoran, Settleman, & Engelman, 2011; Goldinger, Murer, Stieger, & Dummer, 2013). MEK inhibitors, such as trametinib, cobimetinib and binimetinib, work by binding the MEK kinase protein regardless of BRAF mutation status, resulting in

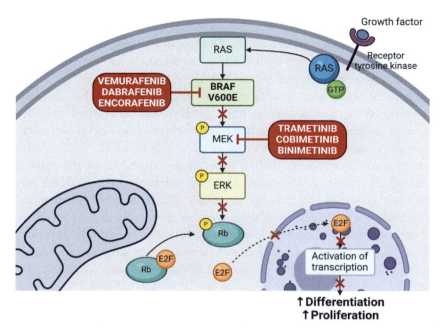

Fig. 1 Altered RAS-RAF-MEK-ERK signaling pathway in BRAF mutant melanoma cells and associated targeted therapies. See main text for details. *Figure created with BioRender.com.*

inhibition of uncontrolled cell growth. Treatment with both BRAF and MEK inhibitors have been employed for late stage melanoma treatment (Schadendorf et al., 2024). The combination therapy doubled progression-free survival at two and three years, compared to just BRAF inhibitor treatment alone (Long et al., 2018).

3. Drug resistance in melanoma

While recent developments in treatment for metastatic melanoma have been effective, a major issue is the development of drug resistance. In patients that show an initial response, the mean duration of response is approximately 6 months, with up to 50% of patients developing resistance to vemurafenib within this period (Chapman et al., 2011). Mechanisms of resistance can be separated according to the dependence on ERK (Corcoran et al., 2011). Alterations to the Ras-Raf-MEK-ERK pathway can result in reactivation of ERK despite the presence of mutant BRAF, while amplification of the mutated *BRAF* gene has also been identified as conferring resistance to vemurafenib (Corcoran et al., 2010). Another study using mouse xenograft models demonstrated that vemurafenib resistant tumors had amplification of the mutant gene and elevated mutant BRAF protein expression (Das Thakur et al., 2013). However, amplification of the mutant gene has not been demonstrated in clinical samples from patients who developed treatment resistance. Increased activity of other Raf proteins such as C-Raf also induce resistance to vemurafenib (Montagut et al., 2008; Su et al., 2012). While alternative mutations of the *BRAF* gene do not appear to be the cause of vemurafenib resistance, additional activating mutations in the *NRAS* gene have been identified (Nazarian et al., 2010). BRAF and NRAS mutations are typically mutually exclusive, with very few melanoma tumors presenting with mutations in both genes (Tran et al., 2021). However, in both resistant mutant BRAF melanoma cell lines and patient tumor samples an activating NRAS mutation in the Q61 amino acid position has been identified. This mutation can lead to reactivation of the Ras-Raf-MEK-ERK signaling pathway through other Raf isoforms.

Alternate resistance mechanisms can enable cell proliferation independent of ERK reactivation (Corcoran et al., 2011). Increased expression of receptor tyrosine kinases is the most common mechanism. Platelet derived growth factor receptor beta (PDGFR-β) receptor tyrosine kinase has many roles in cancer cells but has recently been identified as being upregulated in vemurafenib-resistant melanoma cell lines (Nazarian et al., 2010). Insulin like growth factor 1 receptor (IGF1R) is also upregulated in vemurafenib-resistant melanoma cells (Villanueva et al., 2010). However, in the cases of both PDGFR-β and IGF1R, knockdown of these dominant receptor tyrosine kinases did not result in increased sensitivity to BRAF inhibitors, suggesting that additional signaling pathways are also involved in the acquisition of resistance.

Conventional resistance mechanisms have also been described in BRAF-inhibitor resistant melanoma. ATP-binding cassette (ABC) transporters are known to pump compounds out of cells, contributing to drug resistance (Szakács et al., 2014). In melanoma, ABC transporters have been described as playing a key role in reducing the effect of vemurafenib, leading to limited distribution of vemurafenib in metastatic sites (Wu & Ambudkar, 2014). Additionally, multidrug transporter Patched is strongly expressed in metastatic melanoma tissue and associated with decreased survival (Signetti et al., 2020). Patched inhibition led to increased vemurafenib effectiveness in resistant melanoma cells.

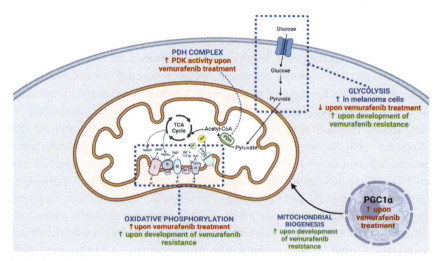

Fig. 2 **Metabolic reprogramming reported in metastatic melanoma cells treated with and resistant to vemurafenib or other BRAF-inhibitors.** See main text for details. *Figure created with BioRender.com.*

The role of metabolic reprogramming in the development of targeted therapy resistance has also been investigated (Fig. 2). Cancer cells are able to preferentially use aerobic glycolysis over mitochondrial oxidative phosphorylation to increase their energy and metabolite production for proliferation (Liberti & Locasale, 2016; Vander Heiden, Cantley, & Thompson, 2009). This Warburg effect has been demonstrated in melanoma cells, with increased levels of glycolysis in normoxic and hypoxic conditions compared to non-cancerous melanocytes (Scott et al., 2011). Interestingly, BRAF mutant melanoma cells have been demonstrated to reduce glucose uptake and switch back to mitochondrial oxidative phosphorylation following vemurafenib treatment (Corazao-Rozas et al., 2016; Haq et al., 2013). This is driven by increased expression of genes involved in the tricarboxlic acid (TCA) cycle and oxidative phosphorylation (Haq et al., 2013). Further investigation showed that melanoma cells increased basal and maximum respiratory capacity after short term exposure to vemurafenib (Corazao-Rozas et al., 2016). Additionally, a shift from glycolysis to mitochondrial oxidative phosphorylation was demonstrated in vemurafenib-reated cells by decreased extracellular acidification rate, glucose uptake and expression of glucose transporters, coupled with an increase in oxygen consumption rate.

Increased mitochondrial oxidative phosphorylation and mitochondrial biogenesis has also been measured in vemurafenib-resistant melanoma cells (Khamari et al., 2018). While glycolysis is reported to be suppressed in response to BRAF inhibitor treatment, recovery occurs in cells that escape BRAF inhibition (Parmenter et al., 2014; Verduzco, Flaherty, & Smalley, 2015). Additionally, increased glutaminolysis has been measured in vemurafenib-resistant melanoma cells. Khamari et al. demonstrated that melanoma cells isolated from mouse xenografts that had developed vemurafenib resistance had increased glutamine consumption that fed into the TCA cycle, therefore increasing mitochondrial oxidative phosphorylation (Khamari et al., 2018). Changes in mitochondrial calcium uniporter expression has also been reported in BRAF inhibitor resistant melanoma cells, with decreased expression resulting in reduced mitochondrial calcium uptake that further caused upregulation of the pentose phosphate pathway and the TCA cycle (Stejerean-Todoran et al., 2022).

The mechanism associated with increased oxidative phosphorylation in vemurafenib-resistant melanoma cells following vemurafenib treatment is reported to be dependent on peroxisome proliferator-activated receptor-gamma coactivator 1 alpha (PGC1α) (Corazao-Rozas et al., 2013; Haq et al., 2013) via increased microphthalmia-associated transcription factor (MITF) activity (Wolf, 2014). PGC1α is a transcriptional coactivator that stimulates mitochondrial biogenesis, and is known as a master regulator of mitochondrial metabolism. Increased PGC1α has been measured in melanoma cells treated with vemurafenib (Haq et al., 2013), along with increased pyruvate dehydrogenase kinase (PDK) activity that regulates pyruvate entry into the TCA cycle (Cesi, Walbrecq, Zimmer, Kreis, & Haan, 2017).

4. Oxidative stress in melanoma

General dogma states that cancer cells are under more oxidative stress than normal cells. Melanoma cells fit that profile, with primary melanocytes themselves exposed to more oxidative stress from a number of external and internal sources than many other cell types (Fig. 3). The most obvious is UV radiation. The majority (95%) of UV radiation is in the form of UVA (Swalwell, Latimer, Haywood, & Birch-Machin, 2012). While UVB radiation is higher energy and can cause direct DNA damage, UVA radiation penetrates the epidermis and dermis more deeply, affecting melanocytes.

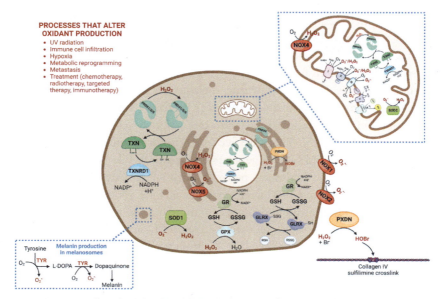

Fig. 3 Sources of antioxidants and oxidants in melanoma cells. See main text for details. *Figure created with BioRender.com.*

UVA radiation leads to molecular excitation and electron transfer reactions, resulting in increased production of reactive oxygen species.

One of the main roles of melanocytes is to produce melanin, a class of pigments synthesized from tyrosine that protect the nuclei of melanocytes and surrounding epidermal cells from UV-induced damage. However, oxidants are also produced as a by-product of melanin synthesis (Kim, Panzella, Napolitano, & Payne, 2020). Tyrosinase is the rate-limiting enzyme in the synthesis of melanin, where it. catalyzes the oxidation of tyrosine to dopamine, and then dehydrogenating dopamine to dopaquinone (Koga, Nakano, & Tero-Kubota, 1992). These steps result in the reduction of oxygen to superoxide (Denat et al., 2014).

When melanin synthesis is inhibited in melanocytes, lower levels of intracellular oxidants are measured (Jenkins & Grossman, 2013). Melanin is typically contained within melanosome structures, protecting the cell from oxidative damage, but disrupted melanosomes are reported in melanoma cells (Gidanian et al., 2008; Kim et al., 2020). There are two forms of melanin, darkly pigmented eumelanin and lightly pigmented sulfur-containing pheomelanin (Sample & He, 2018). Eumelanin is described as being photoprotective, protecting melanocyte nuclei from damage by UV radiation.

In contrast, pheomelanin is phototoxic, generating superoxide when excited by UVA radiation (Kim et al., 2020). Melanin has also been reported to bind to mitochondrial DNA in melanoma cells, further increasing its damaging properties (Swalwell et al., 2012).

Another source of oxidants within melanoma cells are the nicotinamide adenine dinucleotide phosphate (NADPH) oxidases (NOX). These membrane-bound enzyme complexes catalyze one- or two- electron reduction of oxygen (Roy et al., 2015). There are five NOX isoforms and two dual oxidases (DUOX). NOX1–3 and NOX5 generate superoxide, while NOX4, DUOX1 and DUOX2 produce hydrogen peroxide (H_2O_2) directly (Meitzler et al., 2014). Additional oxidant-producing enzymes include xanthine oxidase, glucose oxidase and D-amino acid oxidase. Xanthine oxidase produces superoxide that dismutates to H_2O_2, while glucose oxidase and D-amino acid oxidase both directly produce H_2O_2 (Halliwell & Gutteridge, 2015). H_2O_2-producing NOX4 and superoxide-producing NOX5 have been reported to be highly expressed in melanoma cells at both an mRNA and protein level (Antony et al., 2017; Carvalho et al., 2022; Meitzler, Konaté, & Doroshow, 2019). Additionally, high NOX4 protein expression has been measured in melanoma tumor samples and associated with cell proliferation, epithelial-mesenchymal transition and invasiveness (Meitzler et al., 2017, 2019).

The process of invasion and metastasis promotes oxidative stress in cancer cells. Detachment from the cell matrix has been shown to induce oxidant production in breast cancer cells (Schafer et al., 2009). Furthermore, the vasculature is suggested to be a more oxidative environment due to higher oxygen tensions. This is proposed to be a protective mechanism of the human body, with a more oxidative environment in the vascular system potentially limiting distant metastasis (Piskounova et al., 2015). In a mouse study, melanoma cells collected from blood vessels and metastatic nodules had higher levels of cytosolic and mitochondrial oxidants compared to subcutaneous tumors. Additionally, melanoma cells that metastasized through the lymphatic system had higher rates of survival compared to the vascular system (Ubellacker et al., 2020).

Mitochondria can generate significant amounts of superoxide (Murphy, 2009). The metabolic reprogramming in cancer cells described earlier, in particular increased oxidative phosphorylation, may be associated with increased mitochondrial oxidant production (Gorrini, Harris, & Mak, 2013). While glycolysis is upregulated in cancer cells, melanoma cells undergo a switch back to mitochondrial oxidative phosphorylation when becoming invasive, with an associated increase in oxidant production

(Abildgaard & Guldberg, 2015; Salhi et al., 2020). More research is required, however, to reveal the extent and significance of any observed increases. Electron leakage from the electron transport chain is a demonstrated source of superoxide, but there are at least 11 different oxidant sources in mitochondria, including 2-oxoacid dehydrogenase complexes, flavin-containing site and quinone binding sites of complex I, ubiquinone binding site at complex III, succinate dehydrogenase, electron-transferring flavoproteins and dihydroorate dehydrogenase (Brand, 2016; Quinlan, Perevoshchikova, Hey-Mogensen, Orr, & Brand, 2013). Many of the sites generate superoxide and/or H_2O_2 into the mitochondrial matrix, though some are located on the outer side of the mitochondrial inner membrane and therefore produce oxidants to the intermembrane space in addition to the mitochondrial matrix.

5. Antioxidants in melanoma

Antioxidants protect cells by preventing oxidant production, scavenging oxidants and promoting the repair of oxidative damage (Halliwell & Gutteridge, 2015). Endogenous antioxidant expression is largely regulated by nuclear factor erythroid 2-related factor 2 (NRF2) (Yamamoto, Kensler, & Motohashi, 2018). NRF2 binds to the antioxidant response element (ARE) in the promoter regions of more than 200 genes. As described earlier, in addition to the oxidative challenges faced by other cell types, melanocytes need effective antioxidant systems to cope with UV radiation.

With the transformation of melanocytes to melanoma cells resulting in increased oxidative stress, it is reported that their antioxidants are upregulated to protect the cells from oxidative damage (Fig. 3). There are two major antioxidant networks in cells: the glutathione and thioredoxin systems. The thioredoxin system is an NADPH-dependent system that comprises of thioredoxin (TXN) and thioredoxin reductase (TXNRD) (Lu & Holmgren, 2014). Thioredoxin reduces proteins containing oxidized cysteines and is recycled by NADPH-dependent thioredoxin reductase. There are two mammalian isoforms of thioredoxin: cytoplasmic TXN and mitochondrial TXN2, and three mammalian isoforms of thioredoxin reductase: cytoplasmic TXNRD1, mitochondrial TXNRD2, and TXNRD3, which is a tissue-specific isoform found only in male germ cells.

Components of the thioredoxin system are reported to be upregulated in cancer cells, including melanoma, and play a key role in protecting cancer cells from oxidative stress and apoptosis while promoting proliferation (Gencheva & Arnér, 2022; Holmgren & Lu, 2010). Alterations in the thioredoxin gene were identified in 34% of patients, with a reduction in mRNA expression recorded in metastatic melanoma samples compared to primary melanoma tumors (Carvalho et al., 2022). However, studies in melanoma tissue samples have also demonstrated increased thioredoxin protein expression, which was associated with advanced melanoma progression (Wang, Dong, Li, Li, & Hong, 2015). This is further demonstrated in advanced melanoma tissue samples, which had increased thioredoxin protein expression compared to melanomas of less aggressive nature (Lincoln, Ali Emadi, Tonissen, & Clarke, 2003). Thioredoxin reductase expression is mainly reported to be increased in melanoma, with elevated mRNA expression of both cytosolic *TXNRD1* and mitochondrial *TXNRD2* (Carvalho et al., 2022). Interestingly, higher *TXNRD2* mRNA expression was also associated with a significant reduction in overall melanoma patient survival. Tissue microarrays showed increased TXNRD1 protein expression that was associated with melanoma progression (Cassidy et al., 2015). Another protein involved in the thioredoxin system is the thioredoxin interacting protein (TXNIP), which is an endogenous inhibitor of thioredoxin that competes with other binding proteins of thioredoxin, such as apoptosis signal–regulating kinase 1 (ASK1). Decreased *TXNIP* mRNA expression has been demonstrated in melanoma cells at advanced stages (Carvalho et al., 2022; Meylan et al., 2021).

Peroxiredoxins (PRDX) are a family of thiol–dependent peroxidase enzymes that are highly reactive with H_2O_2 and recycled by the thioredoxin system (Rhee, 2016). Peroxiredoxins are abundantly expressed, making up almost 1% of total protein within the cell. There are six mammalian peroxiredoxins, each occupying different subcellular compartments: PRDX1, PRDX2 and PRDX6 are found in the nucleus and cytosol, PRDX3 in the mitochondria, PRDX4 in the endoplasmic reticulum, and PRDX5 in the cytosol, mitochondria and peroxisomes (Poole & Nelson, 2016). Analysis of genomic data taken from the Cancer Genome Atlas Program revealed *PRDX* genes were altered in 12% of melanoma (Schmitt, Schmitz, Hufnagel, Schartl, & Meierjohann, 2015). *PRDX* mRNA expression has been evaluated in benign nevi, dysplastic nevi and melanomas of different stages of development and altered expression levels appear to be dependent on the stage of melanoma progression (Carvalho et al., 2022). During the vertical growth phase of primary melanoma tumors, increased *PRDX1*, *PRDX4* and *PRDX5* mRNA expression was measured, while *PRDX2* and *PRDX6* mRNA expression was decreased.

Once metastatic, mRNA expression of *PRDX3*, *PRDX4* and *PRDX6* increased, while *PRDX1* and *PRDX5* decreased. Decreased PRDX1 and PRDX2 protein expression was measured in melanoma compared to benign and dysplastic nevi (Hintsala, Soini, Haapasaari, & Karihtala, 2015). In a panel of melanoma cell lines, variable PRDX2 protein expression was measured, with some having very low detectable levels of PRDX2 protein (Lee et al., 2013). Using a swordfish model, PRDX2 and PRDX6 have been demonstrated to be upregulated in malignant melanoma (Lokaj et al., 2009). In a rare form of melanoma, uveal melanoma, increased PRDX3 was associated with metastasis and poor survival (Ramasamy et al., 2020).

The glutathione antioxidant system is based on the tripeptide glutathione (GSH) (Holmgren et al., 2005; Lillig, Berndt, & Holmgren, 2008). In addition to reacting directly with some oxidants and electrophiles, glutathione is a substrate for glutathione peroxidases (GPX) to detoxify hydroperoxides, becoming oxidized to its disulfide (GSSG) and then being recycled by NADPH-dependent glutathione reductase (GR) (Carvalho et al., 2022). Glutaredoxin (GLRX) reduces GSH adducts on protein thiols (Berndt, Lillig, & Holmgren, 2008). In melanoma, *GPX* mRNA expression is increased compared to benign nevi (Carvalho et al., 2022). As melanoma becomes metastatic, *GPX3* and *GLRX* mRNA expression decreases, while *GR* mRNA expression increases. GPX3 protein overexpression was reported to inhibit melanoma cell proliferation (Yi et al., 2019). Inhibition of glutathione synthesis has also been demonstrated to increase sensitivity of melanoma cells to the chemotherapeutic agent, temozolomide, suggesting that glutathione may contribute to the acquisition of resistance (Rocha, Kajitani, Quinet, Fortunato, & Menck, 2016). Mouse studies have also revealed that melanoma cells metastasizing through the blood have increased levels of GSSG and GPX4 along with upregulated NADPH levels, indicative of metastatic melanoma cells being under more oxidative stress (Piskounova et al., 2015; Ubellacker et al., 2020).

Peroxidasin (PXDN), an extracellular matrix peroxidase, is under control of NRF2 (Hanmer & Mavri-Damelin, 2018) and snail family transcriptional repressor 1 (SNAI1), a master transcription regulator of epithelial-mesenchymal transition (EMT) (Sitole & Mavri-Damelin, 2018). Peroxidasin expression is associated with the invasive melanoma phenotype (Paumann-Page et al., 2021) and upregulated during melanoma progression (Carvalho et al., 2022). Knock-down of peroxidasin reduces melanoma cell invasion and promotes a transcriptomic shift towards a less invasive more epithelial-like phenotype (Jayachandran et al., 2016; Smith-Díaz et al., 2024). Peroxidasin is translated and processed in the endoplasmic reticulum and golgi apparatus and subsequently secreted to the extracellular matrix. Peroxidasin uses H_2O_2 to oxidize bromide to form hypobromous acid (HOBr) extracellularly and inside melanoma cells (Paumann-Page et al., 2021). In healthy tissue, peroxidasin is resident in the basement membrane where it binds to laminin and forms covalent crosslinks in collagen IV by HOBr (Bhave et al., 2012). However, laminin and collagen IV expression is dysregulated in many cancers and the role peroxidasin plays in solid tumors is still unclear. Its catalytic activity may protect melanoma cells from oxidative stress by reacting with H_2O_2, as previously shown for prostate cancer (Dougan et al., 2019). However, peroxidasin generates oxidants with different physiological properties and therefore may be acting as a redox switch that protects from H_2O_2-induced cell death and promotes proliferation and invasion. Carvalho et al. have reported significant upregulation of *PXDN* mRNA during the vertical growth phase of melanoma but interestingly not in metastatic melanoma (Carvalho et al., 2022). However, as peroxidasin expression seems phenotype-specific in metastatic melanoma, this result may mask differential peroxidasin expression in a heterogenous population present in a patient tumor sample.

While melanin can be a pro-oxidant, it also has antioxidant properties. Darkly pigmented eumelanin, which is synthesized by melanocortin 1 receptor (MC1R), is described as being photoprotective (Kim et al., 2020), protecting nuclei from UV-induced DNA damage. MC1R has also been reported to induce expression of antioxidants such as catalase through activation with alpha-melanocyte-stimulating hormone (αMSH) (Denat et al., 2014). Melanin also protects mitochondrial DNA from damage induced by H_2O_2 (Swalwell et al., 2012).

Other antioxidants are also reported to have altered expression during melanoma progression, including superoxide dismutases (SOD) (Carvalho et al., 2022). Increased expression of *SOD1* was identified in melanoma cells compared to melanocytic nevi, while expression of *SOD3* and catalase was decreased. In addition to NRF2, forkhead box-O (FOXO) transcription factors have also been implicated in transcription of SOD, catalase and sestrin3. Activating transcription factor 4 (ATF4) is an additional transcription factor that is upregulated in melanoma cells, inducing expression of genes involved in the synthesis of glutathione (Tirosh et al., 2016). Melanoma cells with high levels of PGC-1α, a master regulator of mitochondrial biogenesis, are reported to have an increased antioxidant capacity through SOD2 and thioredoxin, glutaredoxin and peroxiredoxin family members (Vazquez et al., 2013).

6. Oxidative stress and BRAF inhibitor resistance

Metabolic changes associated with the acquisition of drug resistance can result in increased sensitivity to other stressors, a phenomenon termed acquired vulnerability or collateral sensitivity. Early we described the metabolic reprogramming that occurs in drug resistant cells. In this section we discuss what is known about changes in redox homeostasis in melanoma cells during the development of resistance to BRAF inhibitors, and how drug-resistant cells may be more susceptible to treatments that target their antioxidant pathways.

Treatment of melanoma cells with BRAF inhibitors has been reported to induce oxidative stress in melanoma cells, with increases in both cytosolic and mitochondrial oxidants (Cesi et al., 2017; Corazao-Rozas et al., 2013; Guo et al., 2023; Haq et al., 2013). Increased mitochondrial superoxide, as measured by the MitoSOX probe, was detected in melanoma cells after exposure to vemurafenib for 24–72 h (Corazao-Rozas et al., 2013; Haq et al., 2013). Even after vemurafenib was removed from melanoma cells, increased mitochondrial superoxide production was maintained (Corazao-Rozas et al., 2013). Melanoma cells transfected with plasmids encoding cytoplasmic Grx2-roGFP2 or mitochondrial mito-roGFP2 and treated with vemurafenib for 3–10 h displayed increased mitochondrial and cytosolic oxidant production (Cesi et al., 2017). The mito-roGFP2 signal was decreased by mitochondrial-targeted ubiquinone. Cells that had developed resistance to combined treatment with BRAF inhibitor dabrafenib and MEK inhibitor trametinib displayed even further increased cytosolic oxidant production (Wang et al., 2018).

In addition to increased oxidant production in BRAF inhibitor-resistant melanoma, increased expression of antioxidant genes have also been reported. Increased GSH content and oxidized GSSG was measured in BRAF inhibitor resistant melanoma cells (Bishal Paudel et al., 2020; Corazao-Rozas et al., 2013; Khamari et al., 2018), which was associated with increased NRF2 activation (Khamari et al., 2018) and upregulation of catalase (Corazao-Rozas et al., 2013), thioredoxin reductase and sestrin2 (Guo et al., 2023; Yang et al., 2021). Cytosolic peroxiredoxins were also reported to be altered in a vemurafenib-resistant subpopulation of melanoma cells, with PRDX1 expression decreasing and PRDX2 oxidation increasing (Carvalho et al., 2024). Increased production of NADPH, which is critical for the recycling of antioxidants, has been measured in vemurafenib-resistant melanoma cells (Khamari et al., 2018). Furthermore, melanoma cells resistant to BRAF- and MEK-inhibitors have high levels of *AXL* expression, which encodes for tyrosine-protein kinase receptor UFO and is activated by ATF4, a transcription factor that also leads to expression of antioxidant genes (Tirosh et al., 2016). Increased levels of cytoplasmic superoxide have been measured in MEK-inhibitor resistant melanoma cells, in addition to decreased levels of GSH (Eichhoff et al., 2023). On the contrary, another study has demonstrated decreased mitochondrial H_2O_2 in BRAF-inhibitor resistant melanoma cells (Stejerean-Todoran et al., 2022).

These studies suggest that melanoma cells that acquire treatment resistance undergo redox changes that may be critical to their ability to survive, and this provide a target for treating resistant melanoma cells or preventing the development of drug resistance. Most recently, melanoma subpopulations that are inherently resistant to vemurafenib have been reported to be sensitive to H_2O_2 treatment (Carvalho et al., 2024). Studies involving treatment of BRAF- and MEK-inhibitor resistant melanoma cells with elesclomol, a drug that induces mitochondrial oxidant production through targeting the electron transport chain (Nagai et al., 2012), also showed increased sensitivity to cell death (Cierlitza et al., 2015; Corazao-Rozas et al., 2013; Eichhoff et al., 2023). One of these studies reported that melanoma cells taken from a tumor of a patient that had developed vemurafenib resistance were highly sensitive to elesclomol treatment in vitro (Corazao-Rozas et al., 2013). Further studies have assessed the use of the biguanide antidiabetic drugs metformin and phenformin in combination with BRAF inhibitors. Metformin inhibits complex I of the mitochondrial respiratory chain, increasing cellular adenosine monophosphate (AMP) levels and therefore induces AMP-activated protein kinase (AMPK) activity. Contrasting results have been reported. Phenformin and vemurafenib isoform PLX4720 combination treatment effectively delayed the emergence of drug resistance in melanoma cells (Yuan et al., 2013). Furthermore, a combination of the two drugs reduced the numbers of slow-cycling JARID1B-positive melanoma cells, which are speculated to be a potential source of drug resistance. In contrast, a study investigating the combination of metformin and vemurafenib revealed antagonistic effects on vemurafenib-resistant melanoma cells (Niehr et al., 2011).

Other studies have investigated the role of PI3K/Akt/mTOR signaling pathway and its involvement in metabolic reprogramming in BRAF-inhibitor resistant melanoma. The PI3K/Akt/mTOR signaling pathway is largely involved in cancer cell metabolism, promoting glycolysis and inducing NADPH production via the pentose phosphate pathway (Lien, Lyssiotis, & Cantley, 2016). A combination treatment of PI3K and mTOR inhibitors with BRAF inhibitors overcame BRAF inhibitor resistance in melanoma cells (Deng et al., 2012; Tran et al., 2021; Villanueva et al., 2010; Wang et al., 2021). These findings support the proposal that metabolic reprogramming underlies vemurafenib resistance.

The thioredoxin system is an obvious therapeutic target in melanoma cells (Gencheva & Arnér, 2022; Urig & Becker, 2006). Auranofin is the most well studied thioredoxin reductase inhibitor, approved by the FDA in 1985 for treatment of rheumatoid arthritis (Abdalbari & Telleria, 2021). Auranofin is a gold (I)-containing phosphine compound that is an irreversible pan-inhibitor of both cytosolic and mitochondrial isoforms of thioredoxin reductase. The selectivity of auranofin for thioredoxin reductase comes from the high affinity of gold for thiols, with thioredoxin reductase enzymes having a selenocysteine residue in their active site. While auranofin has been demonstrated to induce cytosolic and mitochondrial oxidative stress by inhibiting the two thioredoxin reductase enzymes at sub-micromolar concentrations (Rigobello, Folda, Baldoin, Scutari, & Bindoli, 2005), additional targets of auranofin have also been identified, such as GPX4 (Cheff et al., 2023), nuclear factor kapa-light-chain-enhancer of activated B cells (NF-κB) p100 subunit 2 (NFKB2) and cysteine- and histidine-rich domain-containing protein 1 (CHORDC1) (Saei et al., 2020). Treatment of cancer cells with auranofin causes dysregulation of mitochondrial function as a result of downregulation of oxidative phosphorylation complexes and outer mitochondrial membrane proteins, and an increase in mitochondrial membrane permeability (Chiappetta et al., 2022). It has been shown that treatment of cancer cells with auranofin causes disruption of mitochondrial redox homeostasis, as measured by increased mitochondrial PRDX3 oxidation, that in turn induces Bax/Bak-dependent apoptosis (Cox, Brown, Arner, & Hampton, 2008). Additionally, endoplasmic reticulum stress and DNA replication stress are induced in cells upon treatment with auranofin, leading to activation of apoptotic cell death pathways (Chiappetta et al., 2022; Freire Boullosa et al., 2021). Clinical trials looking at the effectiveness of auranofin against non-small cell lung cancer (NCT01737502), chronic lymphocytic leukemia (NCT01419691) and ovarian cancer (NCT03456700) have been initiated, with results still pending for the first two. A phase II clinical trial reported that five patients with chronic lymphocytic leukemia and one patient with small lymphocytic lymphoma were treated with 6 mg auranofin daily for 28 days followed by an increase to 9 mg doses (Saba et al., 2013). Stable disease was reported as the best response out of the six patients. Furthermore, an increase in oxidant production was measured in patient cancer cell samples, which was followed by an increase in the number of apoptotic cells and a decrease in NRF2 gene expression. Clinical trials assessing auranofin in melanoma patients have not been performed.

TXNRD1 inhibitor (TRi-1) and TXNRD2 inhibitor (TRi-2) have been identified in a high-throughput drug screen as compounds that irreversibly inhibit cytosolic and mitochondrial isoforms of thioredoxin reductase, respectively (Stafford et al., 2018). Initial results showed that cancer cells were more sensitive to cell death by TRi-1 and TRi-2 compared to non-cancerous cells, whereas auranofin effectively killed both cancer and non-cancerous cells at the same concentration. Further chemical proteomics studies revealed that TRi-1 and TRi-2 were more specific inhibitors of cytosolic and mitochondrial thioredoxin reductase, respectively, than auranofin (Sabatier et al., 2021). Additionally, squamous cell carcinoma tumor growth in mouse models was significantly decreased upon treatment with TRi-1 (Stafford et al., 2018).

Much of our current knowledge of the antioxidant defences of melanoma cells is based on expression data. Cellular antioxidant defences involve coordinated networks involving several proteins and small molecules, and rely on the constant production of reducing equivalents. Increased expression of one component of the system does not necessarily provide extra protection, and similarly, inhibition of one component may not have significant impact on total defence capability. More work is required to assess these networks before and after external challenges. Also, additional effort should go into measuring the redox status of sensitive cellular reporters of redox homeostasis and oxidative stress. The peroxiredoxins fulfill this role, with increased accumulation of disulfide-linked intermolecular dimers upon increased production of H_2O_2 or inhibition of the thioredoxin system (Poynton & Hampton, 2014).

In summary, there is evidence that altered redox homeostasis is associated with the acquisition of drug resistance in melanoma cells. However, further work is required to determine the significance of these alterations, and if this can be exploited to target resistant cells. In addition to targeting drug-resistant cells, there would be value in investigating the processes involved in the acquisition of drug resistance. This can be done with cultured melanoma cells in vitro using sublethal concentrations of drug, then gradually increasing the dose to select for resistant cell lines. Direct monitoring of metabolic changes that occur during this period, particularly at a single cell level, would provide insight into the critical events required for the development of resistance. If drug-resistant melanoma cells are more reliant on their anti-oxidant defences for survival, then they may be more sensitive to compounds that disrupt these defences. A better understanding of the underlying mechanisms holds the promise of identifying new strategies for preventing resistance in patients receiving targeted inhibitors.

Acknowledgments

This work was supported by the Health Research Council of New Zealand.

References

Abdalbari, F. H., & Telleria, C. M. (2021). The gold complex auranofin: new perspectives for cancer therapy. *Discover Oncology, 12*(1), 42. https://doi.org/10.1007/s12672-021-00439-0.

Abildgaard, C., & Guldberg, P. (2015). Molecular drivers of cellular metabolic reprogramming in melanoma. *Trends in Molecular Medicine, 21*(3), 164–171. https://doi.org/10.1016/j.molmed.2014.12.007.

Akbani, R., Akdemir, K. C., Aksoy, B. A., Albert, M., Ally, A., Amin, S. B., ... Ayala, B. (2015). Genomic classification of cutaneous melanoma. *Cell, 161*(7), 1681–1696.

Antony, S., Jiang, G., Wu, Y., Meitzler, J. L., Makhlouf, H. R., Haines, D. C., ... Doroshow, J. H. (2017). NADPH oxidase 5 (NOX5)-induced reactive oxygen signaling modulates normoxic HIF-1α and p27(Kip1) expression in malignant melanoma and other human tumors. *Mol Carcinog, 56*(12), 2643–2662. https://doi.org/10.1002/mc.22708.

Berndt, C., Lillig, C. H., & Holmgren, A. (2008). Thioredoxins and glutaredoxins as facilitators of protein folding. *Biochimica et Biophysica Acta (BBA) - Molecular Cell Research, 1783*(4), 641–650. https://doi.org/10.1016/j.bbamcr.2008.02.003.

Bhave, G., Cummings, C. F., Vanacore, R. M., Kumagai-Cresse, C., Ero-Tolliver, I. A., Rafi, M., ... Hudson, B. G. (2012). Peroxidasin forms sulfilimine chemical bonds using hypohalous acids in tissue genesis. *Nature Chemical Biology, 8*(9), 784–790. https://doi.org/10.1038/nchembio.1038.

Bishal Paudel, B., Lewis, J. E., Hardeman, K. N., Hayford, C. E., Robbins, C. J., Stauffer, P. E., ... Quaranta, V. (2020). An integrative gene expression and mathematical flux balance analysis identifies targetable redox vulnerabilities in melanoma cells. *Cancer Research, 80*(20), 4565–4577. https://doi.org/10.1158/0008-5472.Can-19-3588.

Bollag, G., Tsai, J., Zhang, J., Zhang, C., Ibrahim, P., Nolop, K., & Hirth, P. (2012). Vemurafenib: the first drug approved for BRAF-mutant cancer. *Nature reviews Drug discovery, 11*(11), 873–886. https://doi.org/10.1038/nrd3847.

Brand, M. D. (2016). Mitochondrial generation of superoxide and hydrogen peroxide as the source of mitochondrial redox signaling. *Free Radical Biology and Medicine, 100*, 14–31. https://doi.org/10.1016/j.freeradbiomed.2016.04.001.

Carvalho, L. A. C., Queijo, R. G., Baccaro, A. L. B., Siena, Á. D. D., Silva, W. A., Rodrigues, T., & Maria-Engler, S. S. (2022). Redox-related proteins in melanoma progression. *Antioxidants, 11*(3), 438. https://www.mdpi.com/2076-3921/11/3/438.

Carvalho, L. A. d C., Noma, I. H. Y., Uehara, A. H., Siena, Á. D. D., Osaki, L. H., Mori, M. P., ... Maria-Engler, S. S. (2024). Modeling Melanoma Heterogeneity In Vitro: Redox, Resistance and Pigmentation Profiles. *Antioxidants, 13*(5), 555. https://www.mdpi.com/2076-3921/13/5/555.

Cassidy, P. B., Honeggar, M., Poerschke, R. L., White, K., Florell, S. R., ... Moos, P. J. (2015). The role of thioredoxin reductase 1 in melanoma metabolism and metastasis. *Pigment Cell & Melanoma Research, 28*(6), 685–695. https://doi.org/10.1111/pcmr.12398.

Cesi, G., Walbrecq, G., Zimmer, A., Kreis, S., & Haan, C. (2017). ROS production induced by BRAF inhibitor treatment rewires metabolic processes affecting cell growth of melanoma cells. *Molecular Cancer, 16*(1), 102. https://doi.org/10.1186/s12943-017-0667-y.

Chapman, P. B., Hauschild, A., Robert, C., Haanen, J. B., Ascierto, P., Larkin, J., ... McArthur, G. A. (2011). Improved Survival with Vemurafenib in Melanoma with BRAF V600E Mutation. *New England Journal of Medicine, 364*(26), 2507–2516. https://doi.org/10.1056/NEJMoa1103782.

Cheff, D. M., Huang, C., Scholzen, K. C., Gencheva, R., Ronzetti, M. H., Cheng, Q., ... Arnér, E. S. J. (2023). The ferroptosis inducing compounds RSL3 and ML162 are not direct inhibitors of GPX4 but of TXNRD1. *Redox Biology, 62*, 102703. https://doi.org/10.1016/j.redox.2023.102703.

Chiappetta, G., Gamberi, T., Faienza, F., Limaj, X., Rizza, S., Messori, L., ... Vinh, J. (2022). Redox proteome analysis of auranofin exposed ovarian cancer cells (A2780). *Redox Biology, 52*, 102294. https://doi.org/10.1016/j.redox.2022.102294.

Cierlitza, M., Chauvistré, H., Bogeski, I., Zhang, X., Hauschild, A., Herlyn, M., ... Roesch, A. (2015) Mitochondrial oxidative stress as a novel therapeutic target to overcome intrinsic drug resistance in melanoma cell subpopulations. *Experimental Dermatology, 24*(2), 155–157. https://doi.org/10.1111/exd.12613.

Corazao-Rozas, P., Guerreschi, P., André, F., Gabert, P. E., Lancel, S., Dekiouk, S., ... Kluza, J. (2016). Mitochondrial oxidative phosphorylation controls cancer cell's life and death decisions upon exposure to MAPK inhibitors. *Oncotarget, 7*(26), 39473–39485. https://doi.org/10.18632/oncotarget.7790.

Corazao-Rozas, P., Guerreschi, P., Jendoubi, M., André, F., Jonneaux, A., Scalbert, C., ... Marchetti, P. (2013). Mitochondrial oxidative stress is the Achille's heel of melanoma cells resistant to Braf-mutant inhibitor. *Oncotarget, 4*(11), 1986–1998. https://doi.org/10.18632/oncotarget.1420.

Corcoran, R. B., Dias-Santagata, D., Bergethon, K., Iafrate, A. J., Settleman, J., & Engelman, J. A. (2010). BRAF gene amplification can promote acquired resistance to MEK inhibitors in cancer cells harboring the BRAF V600E mutation. *Science Signaling, 3*(149), ra34. https://doi.org/10.1126/scisignal.2001148.

Corcoran, R. B., Settleman, J., & Engelman, J. A. (2011, Apr). Potential therapeutic strategies to overcome acquired resistance to BRAF or MEK inhibitors in BRAF mutant cancers. *Oncotarget, 2*(4), 336–346. https://doi.org/10.18632/oncotarget.262.

Cox, A. G., Brown, K. K., Arner, E. S. J., & Hampton, M. B. (2008). The thioredoxin reductase inhibitor auranofin triggers apoptosis through a Bax/Bak-dependent process that involves peroxiredoxin 3 oxidation. *Biochemical Pharmacology, 76*(9), 1097–1109. https://doi.org/10.1016/j.bcp.2008.08.021.

Damsky, W. E., Rosenbaum, L. E., & Bosenberg, M. (2010). Decoding melanoma metastasis. *Cancers, 3*(1), 126–163. https://doi.org/10.3390/cancers3010126.

Das Thakur, M., Salangsang, F., Landman, A. S., Sellers, W. R., Pryer, N. K., Levesque, M. P., ... Stuart, D. D. (2013). Modelling vemurafenib resistance in melanoma reveals a strategy to forestall drug resistance. *Nature, 494*(7436), 251–255. https://doi.org/10.1038/nature11814.

Davies, H., Bignell, G. R., Cox, C., Stephens, P., Edkins, S., Clegg, S., ... Futreal, P. A. (2002). Mutations of the BRAF gene in human cancer. *Nature, 417*(6892), 949–954. https://doi.org/10.1038/nature00766.

Denat, L., Kadekaro, A. L., Marrot, L., Leachman, S. A., & Abdel-Malek, Z. A. (2014). Melanocytes as instigators and victims of oxidative stress. *Journal of Investigative Dermatology, 134*(6), 1512–1518.

Deng, W., Vashisht Gopal, Y. N., Scott, A., Chen, G., Woodman, S. E., & Davies, M. A. (2012). Role and therapeutic potential of PI3K-mTOR signaling in de novo resistance to BRAF inhibition. *Pigment Cell & Melanoma Research, 25*(2), 248–258. https://doi.org/10.1111/j.1755-148X.2011.00950.x.

Dougan, J., Hawsawi, O., Burton, L. J., Edwards, G., Jones, K., Zou, J., ... Danaher, A. (2019). Proteomics-Metabolomics Combined Approach Identifies Peroxidasin as a Protector against Metabolic and Oxidative Stress in Prostate Cancer. *International journal of molecular sciences, 20*(12), 3046.

Eichhoff, O. M., Stoffel, C. I., Käsler, J., Briker, L., Turko, P., Karsai, G., ... Levesque, M. P. (2023). ROS induction targets persister cancer cells with low metabolic activity in NRAS-mutated melanoma. *Cancer Research, 83*(7), 1128–1146. https://doi.org/10.1158/0008-5472.Can-22-1826.

Flaherty, K. T., Hodi, F. S., & Fisher, D. E. (2012). From genes to drugs: targeted strategies for melanoma. *Nature Reviews Cancer, 12*(5), 349–361. https://doi.org/10.1038/nrc3218.

Freire Boullosa, L., Van Loenhout, J., Flieswasser, T., De Waele, J., Hermans, C., Lambrechts, H., ... Deben, C. (2021). Auranofin reveals therapeutic anticancer potential by triggering distinct molecular cell death mechanisms and innate immunity in mutant p53 non-small cell lung cancer. *Redox Biology, 42*, 101949. https://doi.org/10.1016/j.redox.2021.101949.

Gencheva, R., & Arnér, E. S. J. (2022). Thioredoxin reductase inhibition for cancer therapy. *Annu Rev Pharmacol Toxicol, 62*, 177–196. https://doi.org/10.1146/annurev-pharmtox-052220-102509.

Gidanian, S., Mentelle, M., Meyskens, F. L., Jr, & Farmer, P. J. (2008). Melanosomal damage in normal human melanocytes induced by UVB and metal uptake - a basis for the pro-oxidant state of melanoma. *Photochemistry and Photobiology, 84*(3), 556–564. https://doi.org/10.1111/j.1751-1097.2008.00309.x.

Goldinger, S. M., Murer, C., Stieger, P., & Dummer, R. (2013). Targeted therapy in melanoma–the role of BRAF, RAS and KIT mutations. *European Journal of Cancer Supplements, 11*(2), 92–96.

Gorrini, C., Harris, I. S., & Mak, T. W. (2013, Dec). Modulation of oxidative stress as an anticancer strategy. *Nat Rev Drug Discov, 12*(12), 931–947. https://doi.org/10.1038/nrd4002.

Guo, S., Yue, Q., Wang, S., Wang, H., Ye, Z., Zhang, W., ... Zhu, G. (2023). Sestrin2 contributes to BRAF inhibitor resistance via reducing redox vulnerability of melanoma cells. *Journal of Dermatological Science, 109*(2), 52–60. https://doi.org/10.1016/j.jdermsci.2022.12.007.

Halliwell, B., & Gutteridge, J. M. (2015). *Free radicals in biology and medicine* (5th ed.). Oxford University Press.

Hanmer, K. L., & Mavri-Damelin, D. (2018). Peroxidasin is a novel target of the redox-sensitive transcription factor Nrf2. *Gene, 674*, 104–114. https://doi.org/10.1016/j.gene.2018.06.076.

Haq, R., Shoag, J., Andreu-Perez, P., Yokoyama, S., Edelman, H., Rowe, G. C., ... Kung, A. L. (2013). Oncogenic BRAF regulates oxidative metabolism via PGC1α and MITF. *Cancer Cell, 23*(3), 302–315.

Hintsala, H.-R., Soini, Y., Haapasaari, K.-M., & Karihtala, P. (2015). Dysregulation of redox-state-regulating enzymes in melanocytic skin tumours and the surrounding microenvironment. *Histopathology, 67*(3), 348–357. https://doi.org/10.1111/his.12659.

Hodis, E., Watson, I. R., Kryukov, G. V., Arold, S. T., Imielinski, M., Theurillat, J. P., ... Chin, L. (2012). A landscape of driver mutations in melanoma. *Cell, 150*(2), 251–263. https://doi.org/10.1016/j.cell.2012.06.024.

Holmgren, A., Johansson, C., Berndt, C., Lönn, M., Hudemann, C., & Lillig, C. (2005). Thiol redox control via thioredoxin and glutaredoxin systems. *Biochemical Society Transactions, 33*(6), 1375–1377.

Holmgren, A., & Lu, J. (2010). Thioredoxin and thioredoxin reductase: current research with special reference to human disease. *Biochemical and Biophysical Research Communications, 396*(1), 120–124.

Jayachandran, A., Prithviraj, P., Lo, P.-H., Walkiewicz, M., Anaka, M., Woods, B. L., ... McKeown, S. J. (2016). Identifying and targeting determinants of melanoma cellular invasion. *Oncotarget, 7*(27), 41186–41202. https://doi.org/10.18632/oncotarget.9227.

Jenkins, N. C., & Grossman, D. (2013). Role of melanin in melanocyte dysregulation of reactive oxygen species. *BioMed Research International, 2013*, 908797. https://doi.org/10.1155/2013/908797.

Khamari, R., Trinh, A., Gabert, P. E., Corazao-Rozas, P., Riveros-Cruz, S., Balayssac, S., ... Kluza, J. (2018). Glucose metabolism and NRF2 coordinate the antioxidant response in melanoma resistant to MAPK inhibitors. *Cell Death & Disease, 9*(3), 325. https://doi.org/10.1038/s41419-018-0340-4.

Kim, E., Panzella, L., Napolitano, A., & Payne, G. F. (2020). Redox activities of melanins investigated by electrochemical reverse engineering: implications for their roles in oxidative stress. *Journal of Investigative Dermatology, 140*(3), 537–543. https://doi.org/10.1016/j.jid.2019.09.010.

Koelblinger, P., Thuerigen, O., & Dummer, R. (2018). Development of encorafenib for BRAF-mutated advanced melanoma. *Current Opinion in Oncology, 30*(2), https://journals.lww.com/co-oncology/fulltext/2018/03000/development_of_encorafenib_for_braf_mutated.11.aspx.

Koga, S., Nakano, M., & Tero-Kubota, S. (1992). Generation of superoxide during the enzymatic action of tyrosinase. *Archives of Biochemistry and Biophysics, 292*(2), 570–575. https://doi.org/10.1016/0003-9861(92)90032-R.

Lee, D. J., Kang, D. H., Choi, M., Choi, Y. J., Lee, J. Y., Park, J. H., ... Kang, S. W. (2013). Peroxiredoxin-2 represses melanoma metastasis by increasing E-cadherin/β-catenin complexes in adherens junctions. *Cancer Research, 73*(15), 4744–4757. https://doi.org/10.1158/0008-5472.Can-12-4226.

Liberti, M. V., & Locasale, J. W. (2016). The Warburg Effect: How Does it Benefit Cancer Cells? *Trends in Biochemical Sciences, 41*(3), 211–218. https://doi.org/10.1016/j.tibs.2015.12.001.

Lien, E. C., Lyssiotis, C. A., & Cantley, L. C. (2016). Metabolic reprogramming by the PI3K-Akt-mTOR pathway in cancer. In T. Cramer, & C. A. Schmitt (Eds.). *Metabolism in Cancer* (pp. 39–72). Springer International Publishing. https://doi.org/10.1007/978-3-319-42118-6_3.

Lillig, C. H., Berndt, C., & Holmgren, A. (2008). Glutaredoxin systems. *Biochimica et Biophysica Acta (BBA) - General Subjects, 1780*(11), 1304–1317. https://doi.org/10.1016/j.bbagen.2008.06.003.

Lin, W. M., Luo, S., Muzikansky, A., Lobo, A. Z. C., Tanabe, K. K., Sober, A. J., ... Duncan, L. M. (2015). Outcome of patients with de novo versus nevus-associated melanoma. *Journal of the American Academy of Dermatology, 72*(1), 54–58. https://doi.org/10.1016/j.jaad.2014.09.028.

Lincoln, D. T., Ali Emadi, E. M., Tonissen, K. F., & Clarke, F. M. (2003). The thioredoxin-thioredoxin reductase system: over-expression in human cancer. *Anticancer research, 23*(3B), 2425–2433. http://europepmc.org/abstract/MED/12894524.

Lokaj, K., Meierjohann, S., Schütz, C., Teutschbein, J., Schartl, M., & Sickmann, A. (2009). Quantitative differential proteome analysis in an animal model for human melanoma. *Journal of Proteome Research, 8*(4), 1818–1827. https://doi.org/10.1021/pr800578a.

Long, G. V., Eroglu, Z., Infante, J., Patel, S., Daud, A., Johnson, D. B., ... Flaherty, K. T. (2018). Long-term outcomes in patients with BRAF V600–mutant metastatic melanoma who received dabrafenib combined with trametinib. *Journal of Clinical Oncology, 36*(7), 667–673. https://doi.org/10.1200/jco.2017.74.1025.

Lu, J., & Holmgren, A. (2014). The thioredoxin antioxidant system. *Free Radical Biology and Medicine, 66*, 75–87. https://doi.org/10.1016/j.freeradbiomed.2013.07.036.

Meitzler, J. L., Konaté, M. M., & Doroshow, J. H. (2019). Hydrogen peroxide-producing NADPH oxidases and the promotion of migratory phenotypes in cancer. *Archives of Biochemistry and Biophysics, 675*, 108076.

Meitzler, J. L., Makhlouf, H. R., Antony, S., Wu, Y., Butcher, D., Jiang, G., ... Doroshow, J. H. (2017). Decoding NADPH oxidase 4 expression in human tumors. *Redox Biology, 13*, 182–195. https://doi.org/10.1016/j.redox.2017.05.016.

Meitzler, J. L., Smitha, A., Yongzhong, W., Agnes, J., Han, L., Guojian, J., ... H, D. J. (2014). NADPH oxidases: a perspective on reactive oxygen species production in tumor biology. *Antioxid Redox Signal, 20*(17), 2873–2889. https://doi.org/10.1089/ars.2013.5603.

Meylan, P., Pich, C., Winkler, C., Ginster, S., Mury, L., Sgandurra, M., ... Michalik, L. (2021). Low expression of the PPARγ-regulated gene thioredoxin-interacting protein accompanies human melanoma progression and promotes experimental lung metastases. *Scientific Reports, 11*(1), 7847. https://doi.org/10.1038/s41598-021-86329-5.

Montagut, C., Sharma, S. V., Shioda, T., McDermott, U., Ulman, M., Ulkus, L. E., ... Settleman, J. (2008). Elevated CRAF as a potential mechanism of acquired resistance to BRAF inhibition in melanoma. *Cancer Research, 68*(12), 4853–4861. https://doi.org/10.1158/0008-5472.Can-07-6787.

Murphy, M. P. (2009). How mitochondria produce reactive oxygen species. *Biochemical Journal, 417*(1), 1–13. https://doi.org/10.1042/BJ20081386.

Nagai, M., Vo, N. H., Shin Ogawa, L., Chimmanamada, D., Inoue, T., Chu, J., ... Wada, Y. (2012). The oncology drug elesclomol selectively transports copper to the mitochondria to induce oxidative stress in cancer cells. *Free Radical Biology and Medicine, 52*(10), 2142–2150. https://doi.org/10.1016/j.freeradbiomed.2012.03.017.

Nazarian, R., Shi, H., Wang, Q., Kong, X., Koya, R. C., Lee, H., ... Lo, R. S. (2010). Melanomas acquire resistance to B-RAF(V600E) inhibition by RTK or N-RAS upregulation. *Nature, 468*(7326), 973–977. https://doi.org/10.1038/nature09626.

Niehr, F., von Euw, E., Attar, N., Guo, D., Matsunaga, D., Sazegar, H., ... Ribas, A. (2011). Combination therapy with vemurafenib (PLX4032/RG7204) and metformin in melanoma cell lines with distinct driver mutations. *Journal of Translational Medicine, 9*(1), 76. https://doi.org/10.1186/1479-5876-9-76.

Parmenter, T. J., Kleinschmidt, M., Kinross, K. M., Bond, S. T., Li, J., Kaadige, M. R., ... McArthur, G. A. (2014). Response of BRAF-mutant melanoma to BRAF inhibition is mediated by a network of transcriptional regulators of glycolysis. *Cancer Discovery, 4*(4), 423–433. https://doi.org/10.1158/2159-8290.Cd-13-0440.

Paumann-Page, M., Kienzl, N. F., Motwani, J., Bathish, B., Paton, L. N., Magon, N. J., ... Winterbourn, C. C. (2021). Peroxidasin protein expression and enzymatic activity in metastatic melanoma cell lines are associated with invasive potential. *Redox Biology*, 102090. https://doi.org/10.1016/j.redox.2021.102090.

Piskounova, E., Agathocleous, M., Murphy, M. M., Hu, Z., Huddlestun, S. E., Zhao, Z., ... Morrison, S. J. (2015). Oxidative stress inhibits distant metastasis by human melanoma cells. *Nature, 527,* 186. https://doi.org/10.1038/nature15726.

Poole, L. B., & Nelson, K. J. (2016, Jan). Distribution and features of the six classes of peroxiredoxins. *Molecules and cells, 39*(1), 53–59. https://doi.org/10.14348/molcells.2016.2330.

Poynton, R. A., & Hampton, M. B. (2014). Peroxiredoxins as biomarkers of oxidative stress. *Biochimica et Biophysica Acta (BBA) - General Subjects, 1840*(2), 906–912. https://doi.org/10.1016/j.bbagen.2013.08.001.

Quinlan, C. L., Perevoshchikova, I. V., Hey-Mogensen, M., Orr, A. L., & Brand, M. D. (2013). Sites of reactive oxygen species generation by mitochondria oxidizing different substrates. *Redox Biology, 1*(1), 304–312. https://doi.org/10.1016/j.redox.2013.04.005.

Ramasamy, P., Larkin, A.-M., Linge, A., Tiernan, D., McAree, F., Horgan, N., ... Meleady, P. (2020). PRDX3 is associated with metastasis and poor survival in uveal melanoma. *Journal of Clinical Pathology, 73*(7), 408–412. https://doi.org/10.1136/jclinpath-2019-206173.

Rheault, T. R., Stellwagen, J. C., Adjabeng, G. M., Hornberger, K. R., Petrov, K. G., Waterson, A. G., ... Uehling, D. E. (2013). Discovery of Dabrafenib: A Selective Inhibitor of Raf Kinases with Antitumor Activity against B-Raf-Driven Tumors. *ACS Medicinal Chemistry Letters, 4(3), 358-362.* https://doi.org/10.1021/ml4000063.

Rhee, S. G. (2016). Overview on peroxiredoxin. *Molecules and cells, 39*(1), 1–5. https://doi.org/10.14348/molcells.2016.2368.

Rigobello, M. P., Folda, A., Baldoin, M. C., Scutari, G., & Bindoli, A. (2005). Effect of auranofin on the mitochondrial generation of hydrogen peroxide - role of thioredoxin reductase. *Free Radical Research, 39*(7), 687–695. https://doi.org/10.1080/10715760500135391.

Rocha, C. R., Kajitani, G. S., Quinet, A., Fortunato, R. S., & Menck, C. F. (2016). NRF2 and glutathione are key resistance mediators to temozolomide in glioma and melanoma cells. *Oncotarget, 7*(30), 48081–48092. https://doi.org/10.18632/oncotarget.10129.

Roy, K., Wu, Y., Meitzler, J. L., Juhasz, A., Liu, H., Jiang, G., ... James, H. (2015). NADPH oxidases and cancer. *Clinical Science, 128*(12), 863–875. https://doi.org/10.1042/CS20140542.

Saba, N. S., Ghias, M., Manepalli, R., Schorno, K., Weir, S., Austin, C., ... Wiestner, A. (2013). Auranofin Induces a Reversible In-Vivo Stress Response That Correlates With a Transient Clinical Effect In Patients With Chronic Lymphocytic Leukemia. *Blood, 122*(21), 3819. https://doi.org/10.1182/blood.V122.21.3819.3819.

Sabatier, P., Beusch, C. M., Gencheva, R., Cheng, Q., Zubarev, R., & Arnér, E. S. J. (2021). Comprehensive chemical proteomics analyses reveal that the new TRi-1 and TRi-2 compounds are more specific thioredoxin reductase 1 inhibitors than auranofin. *Redox Biology, 48,* 102184. https://doi.org/10.1016/j.redox.2021.102184.

Saei, A. A., Gullberg, H., Sabatier, P., Beusch, C. M., Johansson, K., Lundgren, B., ... Zubarev, R. A. (2020). Comprehensive chemical proteomics for target deconvolution of the redox active drug auranofin. *Redox Biology, 32,* 101491. https://doi.org/10.1016/j.redox.2020.101491.

Salhi, A., Jordan, A. C., Bochaca, I. I., Izsak, A., Darvishian, F., Houvras, Y., ... Osman, I. (2020). Oxidative phosphorylation promotes primary melanoma invasion. *The American Journal of Pathology, 190*(5), 1108–1117. https://doi.org/10.1016/j.ajpath.2020.01.012.

Sample, A., & He, Y. Y. (2018). Mechanisms and prevention of UV-induced melanoma. *Photodermatology, photoimmunology & photomedicine, 34*(1), 13–24.

Schadendorf, D., Dummer, R., Flaherty, K. T., Robert, C., Arance, A., de Groot, J. W. B., ... Ascierto, P. A. (2024). COLUMBUS 7-year update: A randomized, open-label, phase III trial of encorafenib plus binimetinib versus vemurafenib or encorafenib in patients with BRAF V600E/K-mutant melanoma. *European Journal of Cancer, 204,* 114073. https://doi.org/10.1016/j.ejca.2024.114073.

Schafer, Z. T., Grassian, A. R., Song, L., Jiang, Z., Gerhart-Hines, Z., Irie, H. Y., ... Brugge, J. S. (2009). Antioxidant and oncogene rescue of metabolic defects caused by loss of matrix attachment. *Nature, 461*(7260), 109–113. https://doi.org/10.1038/nature08268.

Schmitt, A., Schmitz, W., Hufnagel, A., Schartl, M., & Meierjohann, S. (2015). Peroxiredoxin 6 triggers melanoma cell growth by increasing arachidonic acid-dependent lipid signalling. *Biochemical Journal, 471*(2), 267–279.

Scott, D. A., Richardson, A. D., Filipp, F. V., Knutzen, C. A., Chiang, G. G., Ronai, Z. e A., ... Smith, J. W. (2011). Comparative metabolic flux profiling of melanoma cell lines: beyond the Warburg effect. *Journal of Biological Chemistry, 286*(49), 42626–42634. https://doi.org/10.1074/jbc.M111.282046.

Signetti, L., Elizarov, N., Simsir, M., Paquet, A., Douguet, D., Labbal, F., ... Mus-Veteau, I. (2020). Inhibition of Patched Drug Efflux Increases Vemurafenib Effectiveness against Resistant BrafV600E Melanoma. *Cancers, 12*(6), 1500. https://www.mdpi.com/2072-6694/12/6/1500.

Sitole, B. N., & Mavri-Damelin, D. (2018). Peroxidasin is regulated by the epithelial-mesenchymal transition master transcription factor Snail. *Gene, 646,* 195–202. https://doi.org/10.1016/j.gene.2018.01.011.

Smith-Díaz, C. C., Kumar, A., Das, A., Pace, P., Chitcholtan, K., Magon, N. J., ... Paumann-Page, M. (2024). Peroxidasin is associated with a mesenchymal-like transcriptional phenotype and promotes invasion in metastatic melanoma. *bioRxiv, 2024,* 2004.2005.588346. https://doi.org/10.1101/2024.04.05.588346.

Smithers, B. M., Saw, R. P. M., Gyorki, D. E., Martin, R. C. W., Atkinson, V., Haydon, A., ... Thompson, J. F. (2021). Contemporary management of locoregionally advanced melanoma in Australia and New Zealand and the role of adjuvant systemic therapy. *ANZ Journal of Surgery, 91*(S2), 3–13. https://doi.org/10.1111/ans.17051.

Stafford, W. C., Peng, X., Olofsson, M. H., Zhang, X., Luci, D. K., Lu, L., ... Arnér, E. S. J. (2018). Irreversible inhibition of cytosolic thioredoxin reductase 1 as a mechanistic basis for anticancer therapy. *Science translational medicine, 10*(428), eaaf7444.

Stejerean-Todoran, I., Zimmermann, K., Gibhardt, C. S., Vultur, A., Ickes, C., Shannan, B., ... Bogeski, I. (2022). *MCU controls melanoma progression through a redox-controlled phenotype switch. EMBO reports, 23* (11), e54746. https://doi.org/10.15252/embr. 202254746.

Su, F., Bradley, W. D., Wang, Q., Yang, H., Xu, L., Higgins, B., ... Bollag, G. (2012). Resistance to selective BRAF inhibition can be mediated by modest upstream pathway activation. *Cancer Research, 72*(4), 969–978. https://doi.org/10.1158/0008-5472.Can-11-1875.

Swalwell, H., Latimer, J., Haywood, R. M., & Birch-Machin, M. A. (2012). Investigating the role of melanin in UVA/UVB-and hydrogen peroxide-induced cellular and mitochondrial ROS production and mitochondrial DNA damage in human melanoma cells. *Free Radical Biology and Medicine, 52*(3), 626–634.

Szakács, G., Hall, M. D., Gottesman, M. M., Boumendjel, A., Kachadourian, R., Day, B. J., ... Di Pietro, A. (2014). Targeting the Achilles Heel of Multidrug-Resistant Cancer by Exploiting the Fitness Cost of Resistance. *Chemical Reviews, 114*(11), 5753–5774. https://doi.org/10.1021/cr4006236.

Tirosh, I., Izar, B., Prakadan, S. M., Wadsworth, M. H., Treacy, D., Trombetta, J. J., ... Garraway, L. A. (2016). Dissecting the multicellular ecosystem of metastatic melanoma by single-cell RNA-seq. *Science, 352*(6282), 189–196. https://doi.org/10.1126/science. aad0501.

Tran, Gimenez, G., Tsai, P., Kolekar, S., Rodger, E. J., Chatterjee, A., Jabed, A., ... Shepherd, P. R. (2021). Genomic and signalling pathway characterization of the NZM panel of melanoma cell lines: A valuable model for studying the impact of genetic diversity in melanoma. *Pigment Cell & Melanoma Research, 34*(1), 136–143. https://doi.org/10.1111/pcmr.12908.

Tran, K. B., Kolekar, S., Jabed, A., Jaynes, P., Shih, J. H., Wang, Q., ... Shepherd, P. R. (2021). Diverse mechanisms activate the PI 3-kinase/mTOR pathway in melanomas: implications for the use of PI 3-kinase inhibitors to overcome resistance to inhibitors of BRAF and MEK. *BMC Cancer, 21*(1), 136. https://doi.org/10.1186/s12885-021-07826-4.

Tsai, J., Lee, J. T., Wang, W., Zhang, J., Cho, H., Mamo, S., ... Bollag, G. (2008). Discovery of a selective inhibitor of oncogenic B-Raf kinase with potent antimelanoma activity. *Proceedings of the National Academy of Sciences, 105*(8), 3041–3046. https://doi.org/10.1073/pnas.0711741105.

Ubellacker, J. M., Tasdogan, A., Ramesh, V., Shen, B., Mitchell, E. C., Martin-Sandoval, M. S., ... Morrison, S. J. (2020). Lymph protects metastasizing melanoma cells from ferroptosis. *Nature, 585*(7823), 113–118. https://doi.org/10.1038/s41586-020-2623-z.

Urig, S., & Becker, K. (2006). On the potential of thioredoxin reductase inhibitors for cancer therapy. *Seminars in Cancer Biology, 16*(6), 452–465. https://doi.org/10.1016/j.semcancer.2006.09.004.

Vander Heiden, M. G., Cantley, L. C., & Thompson, C. B. (2009). Understanding the Warburg effect: the metabolic requirements of cell proliferation. *Science, 324*(5930), 1029–1033. https://doi.org/10.1126/science.1160809.

Vazquez, F., Lim, J.-H., Chim, H., Bhalla, K., Girnun, G., Pierce, K., ... Puigserver, P. (2013). PGC1α expression defines a subset of human melanoma tumors with increased mitochondrial capacity and resistance to oxidative stress. *Cancer Cell, 23*(3), 287–301. https://doi.org/10.1016/j.ccr.2012.11.020.

Verduzco, D., Flaherty, K. T., & Smalley, K. S. (2015). Feeling energetic? New strategies to prevent metabolic reprogramming in melanoma. *Exp Dermatol, 24*(9), 657–658. https://doi.org/10.1111/exd.12763.

Villanueva, J., Vultur, A., Lee, J. T., Somasundaram, R., Fukunaga-Kalabis, M., Cipolla, A. K., ... Kee, D. (2010). Acquired resistance to BRAF inhibitors mediated by a RAF kinase switch in melanoma can be overcome by cotargeting MEK and IGF-1R/PI3K. *Cancer Cell, 18*(6), 683–695.

Wang, B., Zhang, W., Zhang, G., Kwong, L., Lu, H., Tan, J., ... Guo, W. (2021). Targeting mTOR signaling overcomes acquired resistance to combined BRAF and MEK inhibition in BRAF-mutant melanoma. *Oncogene, 40*(37), 5590–5599. https://doi.org/10.1038/s41388-021-01911-5.

Wang, L., de Oliveira, R. L., Huijberts, S., Bosdriesz, E., Pencheva, N., Brunen, D., ... Los-de Vries, G. T. (2018). An acquired vulnerability of drug-resistant melanoma with therapeutic potential. *Cell, 173*(6), 1413–1425.e1414.

Wang, X., Dong, H., Li, Q., Li, Y., & Hong, A. (2015). Thioredoxin induces Tregs to generate an immunotolerant tumor microenvironment in metastatic melanoma. *OncoImmunology, 4*(9), e1027471. https://doi.org/10.1080/2162402X.2015.1027471.

Wolf, Dieter A. (2014). Is reliance on mitochondrial respiration a "chink in the armor" of therapy-resistant cancer? *Cancer Cell, 26*(6), 788–795. https://doi.org/10.1016/j.ccell.2014.10.001.

Wu, C.-P., & V. Ambudkar, S. (2014). The pharmacological impact of ATP-binding cassette drug transporters on vemurafenib-based therapy. *Acta Pharmaceutica Sinica B, 4*(2), 105–111. https://doi.org/10.1016/j.apsb.2013.12.001.

Yamamoto, M., Kensler, T. W., & Motohashi, H. (2018). The KEAP1–NRF2 system: a thiol-based sensor-effector apparatus for maintaining redox homeostasis. *Physiological Reviews, 98*(3), 1169–1203. https://doi.org/10.1152/physrev.00023.2017.

Yang, C., Tian, C., Hoffman, T. E., Jacobsen, N. K., & Spencer, S. L. (2021). Melanoma subpopulations that rapidly escape MAPK pathway inhibition incur DNA damage and rely on stress signalling. *Nature Communications, 12*(1), 1747. https://doi.org/10.1038/s41467-021-21549-x.

Yi, Z., Jiang, L., Zhao, L., Zhou, M., Ni, Y., Yang, Y., ... Kuang, Y. (2019). Glutathione peroxidase 3 (GPX3) suppresses the growth of melanoma cells through reactive oxygen species (ROS)-dependent stabilization of hypoxia-inducible factor 1-α and 2-α. *Journal of cellular biochemistry, 120*(11), 19124–19136.

Yuan, P., Ito, K., Perez-Lorenzo, R., Del Guzzo, C., Lee, J. H., Shen, C.-H., ... Zheng, B. (2013). Phenformin enhances the therapeutic benefit of BRAF V600E inhibition in melanoma. *Proceedings of the National Academy of Sciences, 110*(45), 18226–18231. https://doi.org/10.1073/pnas.1317577110.

Printed and bound by CPI Group (UK) Ltd, Croydon, CR0 4YY
02/12/2024
01798497-0015